Gene Expression Studies
Using Affymetrix Microarrays

CHAPMAN & HALL/CRC
Mathematical and Computational Biology Series

Aims and scope

This series aims to capture new developments and summarize what is known over the whole spectrum of mathematical and computational biology and medicine. It seeks to encourage the integration of mathematical, statistical and computational methods into biology by publishing a broad range of textbooks, reference works and handbooks. The titles included in the series are meant to appeal to students, researchers and professionals in the mathematical, statistical and computational sciences, fundamental biology and bioengineering, as well as interdisciplinary researchers involved in the field. The inclusion of concrete examples and applications, and programming techniques and examples, is highly encouraged.

Series Editors

Alison M. Etheridge
Department of Statistics
University of Oxford

Louis J. Gross
Department of Ecology and Evolutionary Biology
University of Tennessee

Suzanne Lenhart
Department of Mathematics
University of Tennessee

Philip K. Maini
Mathematical Institute
University of Oxford

Shoba Ranganathan
Research Institute of Biotechnology
Macquarie University

Hershel M. Safer
Weizmann Institute of Science
Bioinformatics & Bio Computing

Eberhard O. Voit
The Wallace H. Couter Department of
Biomedical Engineering
Georgia Tech and Emory University

Published Titles

Bioinformatics: A Practical Approach
Shui Qing Ye

Cancer Modelling and Simulation
Luigi Preziosi

Computational Biology: A Statistical Mechanics Perspective
Ralf Blossey

Computational Neuroscience: A Comprehensive Approach
Jianfeng Feng

Data Analysis Tools for DNA Microarrays
Sorin Draghici

Differential Equations and Mathematical Biology
D.S. Jones and B.D. Sleeman

Exactly Solvable Models of Biological Invasion
Sergei V. Petrovskii and Bai-Lian Li

Gene Expression Studies Using Affymetrix Microarrays
Hinrich Göhlmann and Willem Talloen

Handbook of Hidden Markov Models in Bioinformatics
Martin Gollery

Introduction to Bioinformatics
Anna Tramontano

An Introduction to Systems Biology: Design Principles of Biological Circuits
Uri Alon

Kinetic Modelling in Systems Biology
Oleg Demin and Igor Goryanin

Knowledge Discovery in Proteomics
Igor Jurisica and Dennis Wigle

Modeling and Simulation of Capsules and Biological Cells
C. Pozrikidis

Niche Modeling: Predictions from Statistical Distributions
David Stockwell

Normal Mode Analysis: Theory and Applications to Biological and Chemical Systems
Qiang Cui and Ivet Bahar

Pattern Discovery in Bioinformatics: Theory & Algorithms
Laxmi Parida

Spatiotemporal Patterns in Ecology and Epidemiology: Theory, Models, and Simulation
Horst Malchow, Sergei V. Petrovskii, and Ezio Venturino

Stochastic Modelling for Systems Biology
Darren J. Wilkinson

Structural Bioinformatics: An Algorithmic Approach
Forbes J. Burkowski

The Ten Most Wanted Solutions in Protein Bioinformatics
Anna Tramontano

Chapman & Hall/CRC Mathematical and Computational Biology Series

Gene Expression Studies
Using Affymetrix Microarrays

Hinrich Göhlmann
Willem Talloen

CRC Press
Taylor & Francis Group
Boca Raton London New York

CRC Press is an imprint of the
Taylor & Francis Group an **informa** business
A CHAPMAN & HALL BOOK

Chapman & Hall/CRC
Taylor & Francis Group
6000 Broken Sound Parkway NW, Suite 300
Boca Raton, FL 33487-2742

First issued in paperback 2017

© 2009 by Taylor & Francis Group, LLC
Chapman & Hall/CRC is an imprint of Taylor & Francis Group, an Informa business

No claim to original U.S. Government works

ISBN 13: 978-1-138-11231-5 (pbk)
ISBN 13: 978-1-4200-6515-2 (hbk)

Library of Congress Cataloging-in-Publication Data

Göhlmann, Hinrich.
    Gene expression studies using affymetrix microarrays / Hinrich Göhlmann, Willem Talloen.
        p. cm. -- (Mathematical and computational biology series)
    Includes bibliographical references and index.
    ISBN 978-1-4200-6515-2 (hardcover : alk. paper)
    1. DNA microarrays. 2. Gene expression--Research--Methodology. I. Talloen, Willem. II. Title. III. Series.

QP624.5.D726G64 2009
572.8'636--dc22   2009009976

Visit the Taylor & Francis Web site at
http://www.taylorandfrancis.com

and the CRC Press Web site at
http://www.crcpress.com

To Steffi and Kirsten

Contents

List of Figures

List of Tables

List of BioBoxes

List of StatsBoxes

Preface

Microarray technology   Since its beginnings over 15 years ago, microarray technology has matured. Nowadays, scientists can simultaneously measure the activity of whole genomes on a routine basis due to the reproducibility of commercial platforms. With the increasing quality of commercial microarrays, the inferior quality of self-spotted, dual-color arrays became ever more apparent. Furthermore, statisticians demonstrated the necessity to include dye-swap experiments, decreasing cost effectiveness of making arrays. The complicated dye-swap setups demonstrated that one-color experimental designs are simpler and more flexible to use, both from an experimental and statistical point of view. The current use of commercial single-color microarray technology permits scientists to spend less time in optimizing technical aspects but rather focus on using the technology to answer biological questions.

This book focuses on the use of the Affymetrix GeneChip®[1] system for gene expression studies. The Affymetrix platform is one of the most widely adopted single-color microarray technologies in academic labs as well as in clinics and commercial entities. It has become arguably the microarray platform that is most often used in biomedical research. More than 17,000 scientific publications use or review the technology that is the basis for this gene expression analysis platform.

Applications   Besides gene expression profiling, there are numerous other applications: (1) ChIP on chip for investigating protein binding site occupancy, (2) SNP arrays for studying single nucleotide polymorphisms and copy number variations such as deletions or amplifications, (3) exon arrays to search for alternative splicing events, and (4) tiling arrays for identifying novel transcripts that are either coding or non-coding. This book focuses on gene expression profiling, although the platform has several other application areas. Still, many of the concepts described will apply to these other applications as well.

Due to their high-content nature, microarrays can be used to *generate* hypotheses in contrast to the traditional gene-focused research that was more hypothesis-driven. However, one still has to make sure that the observed differences are in fact caused by the intended treatment and not by a technical

---

[1]Registered trademark of Affymetrix, Inc., Santa Clara, California.

artifact or a confounding effect. Since this is difficult or impossible to disentangle afterwards, the trust in the novel findings largely depends on the quality of the experimental design and the execution of the biological experiment.

As the applications of microarray technology become more numerous and diverse, the limited consensus of how to analyze and interpret the data becomes more apparent. Furthermore, by its increasing popularity, also people with little experience in mRNA experiments are starting to use arrays. Particularly less experienced users may stumble into problems and misleading results. This may unfortunately lead to negative judgments about the technology while it was in fact the experimental design. By addressing issues linked to experimental design, laboratory execution and data analysis, we hope that the merits of microarray technology for gene expression studies may become more clear.

Data analysis Statistical algorithms play a crucial role in high-content biotechnologies. Observed intensities will reflect the true gene expression levels more reliably when better preprocessing techniques are applied. Optimal data analysis approaches are also indispensable. The measured genes represent an enormous amount of various molecular characteristics of a sample at the time of RNA extraction. As it is no longer possible to have an overall view of all available data, we are forced to choose certain viewpoints which are closely linked to the choice for a certain statistical method. Moreover, the comprehensive molecular characterization of the sample allows discovery of interesting findings other than the original research question, but at the same time also increases the risk of identifying false positives. More suitable statistics will raise the true discovery rate and decrease the false discovery rate. Datasets currently tend to become larger and more complex, making the choice for a correct analysis approach even more essential.

There are currently several good books [1],[2] and manuals available on microarray data analysis tools such as R[3], GeneSpring®[2, 3] and GeneMaths®[4]. In this book, the focus lies on the theoretical concepts and the rationale underlying the tools. We intend to make the reader aware of important issues that need to be dealt with when looking at microarray data, such as the curse of dimensionality, the multiple testing problem, correlational vs. experimental research, etc.

The eventual results of a data analysis are numerical findings with statistical significance but not necessarily biological relevance. After the data analysis the biologist has to bring in his biological knowledge to interpret the significant findings and to assess which results make sense given the currently available

---

[2]Registered trademark of Agilent Technologies, Inc., Santa Clara, California.
[3]Originally developed by Silicon Genetics, the software is now sold by Agilent Technologies after acquiring Silicon Genetics.
[4]Registered trademark of Applied Maths, BVBA, St. Martens-Latem, Belgium.

biological knowledge. Pathway analysis approaches can help this biological interpretation process.

Given the overwhelming amount of literature on optimal experimental designs, sample collection and preparation, data preprocessing, statistical analysis and biological interpretation, and given the regular occurence of lack of consensus, the choice for a certain analysis method is not always that easy. Converting the huge number of data points into reliable and biologically relevant results is quite a challenge. This book may help to make analysis choices and to adress the challenge of making sense of microarray output.

Interdisciplinary research Studying gene expression with the help of microarray technology is highly interdisciplinary. It lies at a busy intersection of many different research areas. Microarray studies require input from molecular biology, bioinformatics and (bio)statistics to design, carry out and interpret the results of these experiments. The current state of the technology would neither have been possible without the advances in information technology, combinatorial chemistry and photo-lithography. But also physics has played a role in solving the "riddle of the bright mismatches"[5] or attempts to define the boundaries for using the technology for absolute mRNA quantification.

Although molecular biology, bioinformatics and (bio)statistics were always to some extent interconnected, they were surprisingly separated before the genomic revolution in the 1990s[5]. Bioinformaticians created and advanced algorithms mainly to decode and store DNA sequences of many organisms into databases. Their research contained few statistical issues as they were not testing hypotheses, and contained few biological questions as they rarely interpreted the functionality of genes or mapped them in biological pathways. At the same time, molecular biologists were searching for the molecular causes of manifestations of classical biology by applying biochemical techniques. In contrast to other fields in biology like ecology or evolution, statistics were mostly avoided, and were indeed often less necessary as biologists were only searching for obvious patterns on a limited number of potentially relevant molecules. The statisticians, finally, were focusing on other research areas where the interest and demand was higher. Hence, there was no clear need for a strong interaction between these players. This is the main reason why many researchers studying gene expression lack a comprehensive overview.

For most scientists it will be impossible to understand every part to an extent that they can do the whole experiment themselves. Given the fact that the number and availability of biostatisticians experienced in microarray analysis is rather limited, this book aims to give an overview of all the

---

[5]This refers to an article by Naef and Magnasco in which the authors investigate the balance between the number of labels and the binding efficiency. They demonstrate how the oligonucleotide sequence composition has an impact on both binding energies and labeling probability[4].

relevant aspects and creates awareness of the different issues. Besides educating the reader, it also aims to facilitate communication between different experts. Thereby we hope to make biologists especially aware of the different assumptions behind the statistical methods to avoid misuse of analysis techniques.

Book conventions   Throughout the book we will use the term "gene" synonymously with transcript for simplicity reasons. Of course the reader should be aware of the fact that these two are actually completely different things. Strictly speaking, genes do not have exactly known start or end points along the genomic DNA while transcripts do have these defined boundaries. But oftentimes people refer to genes while they are really referring to a gene product such as a messenger RNA or a protein.

As this book is primarily targeted towards users of the Affymetrix platform, we take the liberty to refer to the Affymetrix arrays also simply as microarrays or just arrays.

We use four different types of boxes in the text to provide more background information about topics that some of the readers might not be familiar with. These boxes are also used to give extra information or highlight important issues:

Yellow Boxes contain important information for all readers. They highlight decision points or common pitfalls, or give guidance in the execution of the experiment and the data analysis.

Blue Boxes contain reference information about the datasets that are used in this book or link to specifics about the Affymetrix platform.

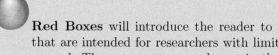**Red Boxes** will introduce the reader to statistical concepts that are intended for researchers with limited statistical background. These concepts are relevant in the context of what is discussed in the text.

Green Boxes will introduce the reader to biological concepts that are intended for researchers with limited background in biology and molecular biology. These concepts are relevant in the context of what is discussed in the text.

All datasets used in this book are publicly available. For the generation of the graphs we have used the program R (version 2.8.2) which is an open source implementation of the programming language S. This language and especially R‘ is one of the main tools currently used for the data analysis of microarray gene expression studies. We are using the R packages bundled together in release 2.3 of the Bioconductor team as well as other publicly available packages.

Every analysis or data graphic in this book has been made using R. We have deliberately not shown or discussed this code in the book to keep the focus on the content. All our code is, however, made available on-line on the book's website, `http://expressionstudies.com`.

Thanks This book describes all of the important experiences in gene expression studies we accumulated over time. As with most sciences, most of these experiences are the result of inspiring discussions with colleagues. This book consequently reflects many insights from our clever fellow data analysts such as Dhammika Amaratunga, Luc Bijnens, Tine Casneuf, Okko Clevert, An De Bondt, Sepp Hochreiter, Steven Osselaer, Nandini Raghavan, Tobias Verbeke, and the team of Ziv Shkedy. But just as important were the thought-provoking questions and comments by insightful scientists as Jeroen Aerssens, Janine Arts, Pieter Peeters, and Hans Winkler, and the excellent lab work of Ilse Van den Wyngaert. Thank you all for your input and the atmosphere of fun and enthusiasm.

Although being generally perceived as adorable women, muses can also appear in the form of male statisticians. Luc Bijnens motivated us to start writing, encouraged and corrected us while writing and brought up excellent ideas for the book.

We also want to thank the cloud. We live in a time where we are privileged to witness the rise of a new paradigm in collecting and exchanging information that will change many things including software development. This book has been made using many open-source products such as R for data analysis, LyX and LaTeX for text editing, JabRef for managing references, and Inkscape and The Gimp for figures. We are therefore indebted to all the people who unselfishly contributed to the current quality of these software tools.

Abbreviations and Terms

AGCC   *Affymetrix GeneChip Command Console* is the name of the current software that operates the whole Affymetrix system.

BGP   *BackGround Probe collection* refers to a background correction technique that is used for, e.g., exon arrays.

Biotin   A molecule that is incorporated during the labeling process to facilitate the attachment of a fluorophore. This dye in turn generates the signal that is proportional to the amount of mRNA present in a sample.

CDF   *Chip Description File* contains layout information about an Affymetrix microarray and the assignment of probe pairs to probesets.

DABG   *Detection Above BackGround* is a detection call generated by comparing perfect match probes to a distribution of background probes.

DNA   DeoxyriboNucleic Acid is the organic molecule that carries the information used by a cell to build the proteins that carry out most of the biological processes in a cell.

FARMS   *Factor Analysis for Robust Microarray Summarization* is a multi-array summarization technique that uses logarithm-transformed data.

FC   *Fold change* for a given gene is the difference between the average expression of two groups in log2 scale.

FDR   *False discovery rate* is an algorithm to estimate the proportion of false positive findings amongst the findings that were called significant.

Feature   A single square on the surface of a microarray chip holding thousands of exact copies of a single type of capture molecule to detect one type of mRNA in a complex sample of different mRNA molecules.

FFPE   *Formalin Fixed Paraffin Embedded* refers to a routine method which is applied to preserve tissue samples for histological evaluations and long-term storage.

GCOS *GeneChip Operating Software* is the name of the former Affymetrix program which controls the array washing stations and the scanner. The current version is called AGCC (see above).

GC-RMA *GC-Robust Multi-array Average* is a multi-array summarization algorithm similar to RMA, but incorporating the MM using a model based on GC content.

Gene A DNA region that typically contains the information on how to build a protein. This information is encoded in a sequence of groups of three bases. The gene sequence can vary in length between 300 bases and hundreds of thousands of bases.

Genome The complete set of all genes in a cell and in a complete organism.

Genotype The genetic makeup of an organism that defines its physical traits (phenotype).

IQR *Inter-quartile range* measures how variable the values in a dataset are, and is calculated as the difference between the 75th and 25th percentile of the data.

IVT *In vitro transcription* is the standard labeling protocol that makes use of an *in vitro* transcription step and a biotin labeling step to generate targets applied to Affymetrix microarrays.

LCM *Laser Capture Microdissection* is a technology that allows the isolation of individual tissue cells by use of a laser through a microscope.

LIMMA *LInear Models for MicroArray data* is a package for the analysis of gene expression microarray data, especially the use of linear models for analysing designed experiments and the assessment of differential expression.

MAQC *MicroArray Quality Control* is an FDA initiated study to assess the reproducibility of microarray results generated by different platforms, and to provide standards for microarray experiments in a clinical setting.

MAS 5.0 *MicroArray Suite 5.0* is a single-array summarization technique introduced by Affymetrix in version 5.0 of their software for the control of array washing and scanning.

MBEI *Model Based Expression Index* is a multi-array summarization technique.

MM *Mismatch* probes are a type of probes on an Affymetrix microarray.

NSMP   *Negative Strand Matching probesets* are derived from Affymetrix's annotation and are used in the PANP detection call algorithm by Peter Warren.

PANP   *Presence-Absence calls with Negative Probesets* is a detection call which uses sets of Affymetrix microarray probes with no known hybridization partners to decide whether a given probeset was detected or not.

PBMC   *Peripheral Blood Mononuclear Cells* are white blood cells playing a role in the immune system.

PLIER   *Probe Logarithmic Intensity ERror* is a multi-array summarization technique minimizing differences in affinity to the biotin-labeled target between the various probes.

PLM   *Probe Level Model* refers to a statistical model that aims to explain mathematically the observed probe signals. Each summarization technique has at its basis such a model that takes into account various effects such as probe affinity, overall chip signal intensity, etc.

PM   *Perfect Match* probes are oligonucleotides on an Affymetrix microarray that capture a single target sequence.

PM-GCBG *Perfect Match* probes are *BackGround* corrected using the median intensity of background probes with the same *GC* content.

Probe   Single-stranded, 25 bases long pieces of DNA immobilized on the surface of the microarray. These oligonucleotides capture transcripts with complementary sequence.

RIN   *RNA Integrity Number* is a number generated from the output data of the Agilent Bioanalyzer which indicates high quality RNA for numbers approaching a maximum value of 10.

RMA   *Robust Multi-array Average* is a multi-array summarization technique.

RT-PCR   *Reverse Transcription Polymerase Chain Reaction* is a molecular biology technique to amplify RNA molecules while the regular polymerase chain reaction was developed to amplify DNA molecules.

SAM   *Significance Analysis of Microarrays* is a statistical technique for finding significant genes in an experiment.

Target   Labeled material synthesized from mRNA that includes fluorescent molecules to quantify the number and abundance of different mRNA species present in the original sample.

VSN   *Variance Stabilizing Normalization* is a method to preprocess microarray intensity data.

Washing   Processing step in the microarray experiment that removes unspecifically bound targets. This leads to an increase in specificity and a decrease in background fluorescent signal.

WT   *Whole Transcript sense target labeling assay* is the laboratory protocol by which the whole transcript is labeled, e.g., for the new Exon and Gene arrays.

Chapter 1

Biological question

All experimental work starts in principle with a question. This also applies to the field of molecular biology. A molecular scientist is using a certain technique to answer a specific question such as, "Does the cell produce more of a given protein when treated in a certain way?" Questions in molecular biology are indeed regularly focused on specific proteins or genes, often because the applied technique cannot measure more.

Gene expression studies that make use of microarrays also start with a biological question. The largest difference to many other molecular biology approaches is, however, the type of question that is being asked. Scientists will typically not run arrays to find out whether the expression of a specific messenger RNA is altered in a certain condition. More often they will focus their question on the treatment or the condition of interest. Centering the question on a biological phenomenon or a treatment has the advantage of allowing the researcher to discover hitherto unknown alterations. On the other hand, it poses the problem that one needs to define when an "interesting" alteration occurs.

1.1 Why gene expression?

1.1.1 Biotechnological advancements

Research evolves and advances not only through the compilation of knowledge but also through the development of new technologies. Traditionally, researchers were able to measure only a relatively small number of genes at a time. The emergence of microarrays (see BioBox 1.1) now allows scientists to analyze the expression of many genes in a single experiment quickly and efficiently.

1.1.2 Biological relevance

Living organisms contain information on how to develop its form and structure and how to build the tools that are responsible for all biological processes that need to be carried out by the organism. This information – the genetic

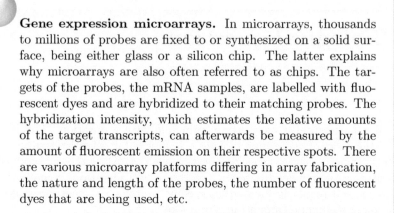

Gene expression microarrays. In microarrays, thousands to millions of probes are fixed to or synthesized on a solid surface, being either glass or a silicon chip. The latter explains why microarrays are also often referred to as chips. The targets of the probes, the mRNA samples, are labelled with fluorescent dyes and are hybridized to their matching probes. The hybridization intensity, which estimates the relative amounts of the target transcripts, can afterwards be measured by the amount of fluorescent emission on their respective spots. There are various microarray platforms differing in array fabrication, the nature and length of the probes, the number of fluorescent dyes that are being used, etc.

BioBox 1.1: Gene expression microarrays

content – is encoded in information units referred to as genes. The whole set of genes of an organism is referred to as its genome.

The vast majority of genomes are encoded in the sequence of chemical building blocks made from deoxyribonucleic acid (DNA) and a smaller number of genomes are composed of ribonucleic acid (RNA), e.g., for certain types of viruses. The genetic information is encoded in a specific sequence made from four different nucleotide bases: adenine, cytosine, guanine and thymine. A slighlty different composition of building blocks is present in mRNA where the base thymine is replaced by uracil. Genetic information encoding the building plan for proteins is transferred from DNA to mRNA to proteins. The gene sequence can range in length typically between hundreds and thousands of nucleotides up to even millions of bases. The number of genes that contain protein-coding information is expected to be between 25,000 to 30,000 when looking at the human genome. A protein is made by constructing a string of protein building blocks (amino acids). The order of the amino acids in a protein matches the sequence of the nucleotides in the gene. In other words, messenger RNA interconnects DNA and protein, and also has some important practical advantages compared to both DNA and proteins (see BioBox 1.2). Increasing our knowlegde about the dynamics of the genome as manifested in the alterations in gene expression of a cell upon treatment, disease, development or other external stimuli, should enable us to transform this knowledge into better tools for the diagnosis and treatment of diseases.

DNA is made of two strands forming together a chemical structure that is called "double helix." The two strands are connected with one another via pairs of bases that form hydrogen bonds between both strands. Such pairing of so-called "complementary" bases occurs only between certain pairs.

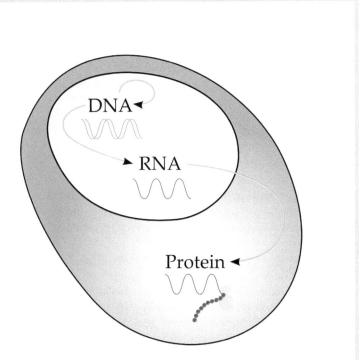

Central dogma of molecular biology. The dogma of molecular biology explains how the information to build proteins is transferred in living organisms. The general flow of biological information (green arrows) has three major components: (1) DNA to DNA (replication) occurs in the cell nucleus (drawn in yellow) prior to cell division, (2) DNA to mRNA (transcription) takes place whenever the cell (drawn in light red) needs to make a protein (drawn as chain of red dots), and (3) mRNA to proteins (translation) is the actual protein synthesis step in the ribosomes (drawn in green). Besides these general transfers that occur normally in most cells, there are also some special information transfers that are known to occur in some viruses or in a laboratory experimental setting.

BioBox 1.2: Central dogma of molecular biology

Hydrogen bonds can be formed between cytosine and guanine or between adenine and thymine. The pairing of the two strands occurs in a process called "hybridization."

Compared to DNA, mRNA is more dynamic and less redundant. The information that is encoded in the DNA is made available for processing in a step called "gene expression" or "transcription." Gene expression is a highly complex and tightly regulated process by which a working copy of the original sequence information is made. This allows a cell to respond dynamically both to environmental stimuli and to its own changing needs, while DNA is relatively invariable. Furthermore, as mRNA constitutes only the expressed part of the DNA, it focuses more directly on processes underlying biological activity. This filtering is convenient as the functionality of most DNA sequences is irrelevant for the study at hand.

Compared to proteins, mRNA is much more measurable. Proteins are 3D conglomerates of multiple molecules and cannot benefit from the hybridising nature of the base pairs in the 2D, single molecule, structure of mRNA and DNA. Furthermore, proteins are very unstable due to denaturation, and cannot be preserved even with very laborious methods for sample extraction and storage.

When using microarrays to study alterations in gene expression, people normally will only want to study the types of RNA that code for proteins – the messenger RNA (mRNA). It is however important to keep in mind that RNA not only contains mRNA – a copy of a section of the genomic DNA carrying the information of how to build proteins. Besides the code for the synthesis of ribosomal RNA, there are other non-coding genes that, e.g., contain information for the synthesis of RNA molecules. These RNAs have different functions that range from enzymatic activities to regulating transcription of mRNAs and translation of mRNA sequences to proteins. The numbers of these functional RNAs that are encoded in the genome are not known. Initial studies looking at the overall transcriptional activity along the DNA are predicting that the number will most likely be larger than the number of protein-coding genes.

People used to say that a large portion of the genomic information encoded in the DNA are useless ("junk DNA"). Over the last years scientific evidence has accumulated that a large proportion of the genome is being transcribed into RNAs of which a small portion constitutes messenger RNAs. All these other non-coding RNAs are divided into two main groups depending on their size. While short RNAs are defined to have sizes below 200 bases, the long RNAs are thought to be mere precursors for the generation of small RNAs, of which the function is currently still unknown – in contrast to the known small RNAs such as microRNAs or siRNAs[6] (see BioBox 1.3 for an overview of different types of RNA). Microarrays are also being made to study differences in abundance of these kinds of RNA.

RNA. In contrast to mRNA (messenger RNA) which contains the information of how to assemble a protein, there are also different types of non-coding RNA (sometimes abbreviated as ncRNA)[a]. Here are the types that are most relevant in the context of this book:

miRNA in length, which regulate gene expression.

long ncRNA (long non-coding RNA) are long RNA molecules that perform regulatory roles. An example is XIST, which can also be used for data quality control to identify the gender of a subject (see BioBox 3.5).

rRNA (ribosomal RNA) are long RNA molecules that make up the central component of the ribosome[b]. They are responsible for decoding mRNA into amino acids and are used for RNA quality control purposes (see Section 3.1.2.8).

siRNA (small interfering RNA) are small double-stranded RNA molecules of about 20-25 nucleotides in length and play a variety of roles in biology. The most commonly known function is a process called RNA interference (RNAi). In this process siRNAs interfere with the expression of a specific gene, leading to a downregulation of the synthesis of new protein encoded by that gene[c].

tRNA (transfer RNA) are small single-stranded RNA molecules of about 74-95 nucleotides in lenghts, which transfer a single amino acid to a growing polypeptide chain at the ribosomal site of protein synthesis. Each type of tRNA molecule can be attached to only one type of amino acid.

---

[a]Non-coding RNA refers to RNA molecules that are transcribed from DNA but not translated into protein.

[b]Ribosomes can be seen as the protein manufacturing machinery of all living cells.

[c]There are, however, also processes known as small RNA-induced gene activation whereby double-stranded RNAs target gene promoters to induce transcriptional activation of associated genes.

BioBox 1.3: siRNA

In this book we will focus on studying mRNA. However, most likely many remarks given on the experimental design and the data analysis will apply to the study of small RNA as well.

1.2   Research question

The key to optimal data analysis lies in a clear formulation of the research question. Being aware of having to define what one considers to be a "relevant" finding in the data analysis step will help in asking the right question and in designing the experiment properly so that the question can really be answered. A well-thought-out and focused research question leads directly into hypotheses, which are both testable and measurable by proposed experiments. Furthermore, a well-formulated hypothesis helps to choose the most appropriate test statistic out of the plethora of available statistical procedures and helps to set up the design of the study in a carefully considered manner. To formulate the right question, one needs to disentangle the research topic into testable hypotheses and to put it in a wider framework to reflect on potentially confounding factors.

Some of the most commonly used study designs in microarray research will be introduced here by means of real-life examples. For each type of study, research questions are formulated and example datasets described. These datasets will be used troughout the book to illustrate some technical and statistical issues.

1.2.1   Correlational vs. experimental research

Microarray research can either be correlational or experimental. In correlational research, scientists generally do not apply a treatment or stimulus to provoke an effect on, e.g., gene expression (influence variables), but measure them and look for correlations with mRNA (see StatsBox 1.1). A typical example are cohort studies, where individuals of populations with specific characteristics (like diseased patients and healthy controls) are sampled and analysed. In experimental research, scientists manipulate certain variables (e.g., apply a compound to a cell line) and then measure the effects of this manipulation on mRNA. Experiments are designed studies where individuals are assigned to specifically chosen conditions, and mRNA is afterwards collected and compared.

It is important to comprehend that only experimental data can conclusively demonstrate *causal* relations between variables. For example, if we found that a certain treatment A affects the expression levels of gene X, then we can conclude that treatment A influences the expression of gene X. Data from

The two interpretations of correlation. Correlation can be defined in a broad and in a narrow sense.

In a **broad sense**, it refers to correlational research. Correlational research involves collecting data in order to determine whether, and to what degree, a relationship exists between two or more variables. An example of correlational research question is, "Is poverty related to success of students in school?" Both poverty and success at school are not necessarily **continuous** variables (like income in dollars as a measure of poverty), but may be **categoric** (like poor/rich). As the relationship is therefore not necessarily linear, its strength will not always be evaluated by correlation coefficients. In the case of a binary variable as poor/rich, for example, it will be evaluated by a t-test.

In a **narrow sense**, correlation refers to the strength and direction of a linear relationship between two, typically continuous, variables.

StatsBox 1.1: The interpretation of correlation

correlational research can only be "interpreted" in causal terms based on some theories that we have, but they cannot conclusively prove causality. If some genes are differentially expressed in a certain disease, we cannot infer from this study whether this differential gene expression is cause or consequence, or whether it is caused by the disease or by some other unmeasured variable related to disease status. The major reason for this is that experimental studies involve a random allocation of subjects to the different conditions to mitigate potential confounding effects, e.g., having only male subjects as controls and only female subjects in the diseased group.

In experiments, the design is directly linked to the research question. As microarrays are typically used for hypothesis generation (what happens in a cell upon a change in conditions?) it is important to have a biological experiment as clean as possible. To ensure this it is advisable to control the outcome of the biological experiment via a different, unrelated measurement and to randomly allocate the subjects to the different experimental conditions. Section 3.1.4.1 discusses the need for this control in more detail.

Some scientific questions

- What genes are differentially expressed between any two groups of the experiment? ⇒ 1.3.1

- Which genes change over time due to a certain treatment or in the cause of a certain process? $\Rightarrow$ 1.3.5

- What are the differences or commonalities between my treatments? $\Rightarrow$ 1.3.4

- What set of genes is predictive for assigning a new sample to a certain class? $\Rightarrow$ 1.3.6

- Which gene expression correlates best with a certain clinical characteristic? $\Rightarrow$ 1.3.6

1.3   Main types of research questions

1.3.1   Comparing two groups

The most widespread research question is the identification of changes in gene activity between two treatment groups, e.g., comparing a treatment or a disease state to a control. The strategies presented here are, however, also applicable to multiple-group-experiments, as biological questions can often easily be broken down into such two-group-comparisons. If one for example looks at data obtained from an experiment with knock-out and wildtype animals profiled in various brain regions with and without compound treatment, one may want to look just at the differences induced by the gene knock-out, the differences induced by the compound in the knock-out animal in the most relevant brain region, etc. All those questions are again two-group comparisons that can be addressed with the strategies described in this section. The effects of multiple comparisons in a single study on the false positive rate will be adressed in Section 5.5.3.

Practical tip: When complex experiments with many sample groups are finally reduced to only a few relevant groups, one should really question the design of the experiment. While complex experiments are justified in certain situations (see Section 1.3.2), it is sometimes preferable to run a simple pilot experiment to first find the relevant brain region, afterwards study the effect of a gene-knock-out in this region, and finally – in the real experiment with a higher sample size – study the effect of the compound in the relevant brain region in the knock-out animals.

Leukemia (Dataset ALL)   A dataset commonly used to illustrate statistical approaches for comparing two groups[1],[7] is the Acute Lymphoblastic Leukemia dataset from the Ritz Laboratory which is available in the R package ALL[8].

Gene expression profile of adult T-cell acute lymphocytic leukemia identifies distinct subsets of patients with different response to therapy and survival.
Sabina Chiaretti, Xiaochun Li, Robert Gentleman, Antonella Vitale, Marco Vignetti, Franco Mandelli, Jerome Ritz, and Robin Foa, *Blood* 2004, 103(7): 2771-2778.

The data consist of microarrays from 128 different individuals with acute lymphoblastic leukemia (ALL). The data were normalized using quantile normalization and summarized using RMA[9]. We will use this dataset to compare the 37 patients with the BCR/ABL fusion gene to the 42 patients from the NEG group, and ignore other available samples and additional covariates.

1.3.2 Comparing multiple groups

When there are several conditions or treatments being potentially interesting, a focus limited to two groups bares the risk of missing interesting patterns. In such a situation, two groups are insufficient for the purposes of the study and it is advisable to include multiple groups.

Practical tip: Regularly, larger studies including multiple groups are planned over time and broken down in several smaller experiments. Although this may help to spread laboratory work and budgets more equally over time, it is unfortunate because of two reasons. First, the use of the same reference group in all separate studies (i.e., a common reference), necessary to allow valid comparisons across studies, inflates the total sample size. Second, comparisons between groups across different studies are difficult to interpret, even when using a common reference, as the studies may differ in some uncontrolled ambient conditions increasing the noise on the signal. Before starting a study, it is therefore wise to reflect on extra, potentially relevant, conditions to include. This makes the study more complete and allows an elegant and simple analysis approach. As mentioned previously, a multiple group study easily allows pairwise comparisons being made between any pair of groups.

Compound profiling (Dataset CP) Compound profiling experiments are commonly applied in pharmaceutical research, as they can help to discover differences between compounds in their mechanism of action. To illustrate this type of research question, we use the data of Dillman et al. (2006)[10] as an example.

TABLE 1.1:   Design two-way
ANOVA.

treatment	wildtype	knock-out
control	Group 1	Group 2
compound	Group 3	Group 4

 Microarray Analysis of Mouse Ear Tissue Exposed to Bis-(2-chloroethyl) Sulfide: Gene Expression Profiles Correlate with Treatment Efficacy and an Established Clinical Endpoint.
J. Dillman, A. I. Hege, C. Phillips, L. Orzolek, A. Sylvester, C. Bossone, C. Henemyre-Harris, R. Kiser, Y. Choi, J. Schlager and C. Sabourin, *JPET* 2006, 317: 76-87.

Dillman et al. (2006) compare three treatment compounds that have been shown to limit sulfur mustard (SM) damage in the mouse ear model: dimercaprol (DM), octyl homovanillamide (OHV) and indomethacin (IM). With microarrays they determined gene expression profiles of biopsies taken from mouse ears after exposure to SM in the presence or absence of treatment compounds. In the examples we limit our analysis to the gene expression measurements of the three compounds in the absence of SM.

1.3.3   Comparing treatment combinations

Many biological activity patterns are condition- or treatment-dependent. Investigating group differences under different conditions or treatments therefore enhances the understanding of the nature of the group differences. Furthermore, it increases the generalizability of the conclusions of the study. Unfortunately, gene expression changes between groups are often studied in only one particular condition, so that one looks at the gene expression in only one way. Including an extra condition or treatment allows one to study gene expression changes in two ways, i.e., changes due to different conditions/treatments as well due to different groups.

Microarray experiments aimed to study treatment combinations need to have two different types of treatments which have been combined in all possible ways. If a researcher for example wants to know whether the effect of a certain compound treatment on mice changes when a gene related to the compound's drug target is knocked-out, he needs to set up an experiment with at least four different groups. He should have wildtype and knock-out mice that are both treated with the compound as well with a control. This results in a design as shown in Table 1.1.

In addition, one can study whether the group differences are indeed condition- or treatment-dependent. Statisticians refer to this as interaction, as the group differences change across conditions. In the example, a significant interaction between knock-out status and treatment will indicate that the effect of the compound significantly changes when a gene related to the compound's drug target is knocked-out.

If the group differences are similar in all conditions, one infers group and condition have additive effects on gene expression levels. More details on treatment combinations and interactions can be found in Section 5.5.2.5.

Glucosamine effects under IL1-β stimulation (Dataset GLUCO)

Exogenous glucosamine globally protects chondrocytes from the arthritogenic effects of IL-1β.
J. Gouze, E. Gouze, M. Popp, M. Bush, E. Dacanay, J. Kay, P. Levings, K. Patel, J. Saran, R. Watson and S. Ghivizzani, *Arthritis Research & Therapy* 2006, 8(6): R173.

Gouze et al. [11] studied the effects of glucosamine on the biology of articular chondrocytes by examining global transcription patterns under normal culture conditions and following challenge with IL1-β. Chondrocytes, isolated from the rats and cultured in flasks, were either alone or in the presence of 20 mM glucosamine. Six hours later, half of the cultures of each group were challenged with 10 ng/ml IL1-β. Fourteen hours after this challenge, RNA was extracted from each culture individually and profiled on rat microarray chips.

Huntington's Disease (Dataset HD)

Brain gene expression correlates with changes in behavior in the R6/1 mouse model of Huntington's disease Genes.
A. Hodges, G. Hughes, S. Brooks, L. Elliston, P. Holmans, S. B. Dunnett, L. Jones, *Brain and Behavior* 2007, 7: 288-299.

The authors of the article use a mouse model to gain mechanistic understanding of Huntington's disease (HD), a disease which causes neurodegeneration leading to an early death. The study was performed to validate the model

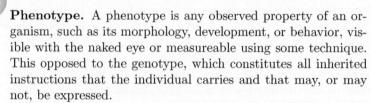

Phenotype. A phenotype is any observed property of an organism, such as its morphology, development, or behavior, visible with the naked eye or measureable using some technique. This opposed to the genotype, which constitutes all inherited instructions that the individual carries and that may, or may not, be expressed.

Almost all molecules except DNA are therefore a part of the phenotype, even when invisible to the naked eye. People regularly refer to the molecular makeup which excludes the genetic information as the "molecular phenotype." Unfortunately, many scientists frequently use phenotype to only refer to externally visible characteristics and forget about the gene products on a molecular level inside the individual cells of an organism.

BioBox 1.4: Phenotype

which uses mice with mutant forms of the huntington gene. Gene expression changes were investigated in a specific brain region of the mouse brain, the striatum, as earlier reports had described an effect on transcription. Three 3-month-old and three 18-month-old mutant mice and six age-matched control mice were studied at time points that coincide with the development of phenotypic alterations (see BioBox 1.4). The main conclusions from the analysis of the data were the identification of affected pathways related to intracellular and electrical signaling, lipid biosynthesis and RNA processes. Furthermore, especially one behavioral change observed in the mice could be correlated with different subsets of gene expression changes[12].

1.3.4   Comparing multiple groups with a reference group

This research question is a special case of comparing multiple groups. The study design also contains more than two groups, but not all pairwise comparisons between all the groups are of interest. The study only wants to compare all the groups against one "Reference" group.

Compound profiling   Compound profiling experiments are commonly applied in pharmaceutical research, as they can help to explore which compounds show a biological activity in certain tissues. This is most commonly done by comparing the compounds under study to the vehicle (i.e., the control). To illustrate this type of research question, we use again the data of Dillman et al. (2006)[10] (see Section 1.3.2) as an example.

Dillman et al. (2006) measured gene expression of biopsies taken from mouse ears after treatment with three compounds and a vehicle. The three compounds (DM, OHV and IM) were administered to the animals in ethanol, and treatment with ethanol (i.e., the vehicle) was also included in the study to check for effects of the vehicle. Here, we limit our analysis to the comparisons of gene expression levels of each of the three compounds vs. the vehicle in the absence of SM.

1.3.5 Investigating within-subject changes

In some microarray experiments the subjects from the different groups are the same. That is, the same subjects have repeatedly been measured, often with some intervention taking place between measures. Using the same subjects has the advantage that subtle treatment or time effects are easier to detect as one can focus on within-individual changes. In other words, the naturally occuring variability between animals can be removed in these studies, as one can assume that the variability between individuals is constant across the repeated measurements.

Such time series experiments are very helpful to discover interesting phenomena, as many biological mechanisms have distinct time profiles. Many time-dependent changes in gene expression may however be irrelevant for the study at hand. Studies using a high number of time points and/or a long duration between time points may therefore pick up very many genes that change over time but are unrelated to a treatment. Such a study will benefit greatly from the inclusion of an appropriate control.

Heat Stress (Dataset HS)

Full genome gene expression analysis of the heat stress response in Drosophila melanogaster.
Jesper G. Sørensen, Morten M. Nielsen, Mogens Kruhøffer, Just Justesen and Volker Loeschcke, *Cell Stress Chaperones* 2005, 10(4): 312–328.

Sørensen et al. (2005)[13] studied changes in gene expression over time induced by a short-term heat treatment in *Drosophila melanogaster* females, using Affymetrix whole genome gene expression arrays. The study focuses on effects just before and at 8 time points after an application of short heat hardening (36°C for 1 hour). The expression changes were followed up to 64 hours after the heat stress, using 4 biological replicates[1].

1.3.6   Classifying and predicting samples

Whenever experiments are done to collect data from samples belonging to two or more groups and the aim is to use a classification algorithm to distinguish between the different groups, it is important to think about a fundamental difference in why one would apply a classification: is it done to describe the class by saying which genes are characteristic for a given class (also called "class description," "class discovery" or "class modeling") or if the experiment is done to define a set of genes that will help in assigning a new sample to a class (also called "class prediction").

Signatures for outcome prognosis (Dataset OP)   The aim of personalized medicine is to treat people according to their individual characteristics. Gene expression profiling can be used as a tool to discover a signature that correlates with this outcome of interest. Such a marker assesses the expression levels of a set of genes. Applying the signature allows to divide people based on their predicted outcome and inform doctors on the most optimal treatment choice for a patient.

Emmprin and Survivin Predict Response and Survival following Cisplatin-Containing Chemotherapy in Patients with Advanced Bladder Cancer.
Als AB, Dyrskjøt L, von der Maase H, Koed K, Mansilla F, Toldbod HE, Jensen JL, Ulhøi BP, Sengeløv L, Jensen KM, Orntoft TF, *Clin Cancer Res.* 2007, 13(15): 4407-4414.

Even though approximately 50% of patients with advanced bladder cancer respond to a chemotherapeutic treatment which contains cisplatin as an active agent, it is still desireable to better predict which patients will benefit from the treatment and how much a beneficial effect will impact the overall survival of responsive patients. In this dataset the authors looked at tumor samples from 30 patients using Affymetrix microarrays. Genes correlating with survival time were identified of which two were validated further. The expression levels of these genes (Emmprin and Survivin) were shown to have strong prognostic value for both response to the treatment as well as survival time[14].

---

[1]In contrast to technical replicates where material from the same sample is profiled two or multiple times, a biological replicate refers to a subject that received the same treatment.

Leukemia (Dataset ALL)   Most classification studies are applied on data from two groups. To illustrate such an approach, we will again use the ALL dataset previously described in Section 1.3.1. For the purpose of this exercise, we will try to classify the 79 patients into their correct genetic group, being the BCR/ABL fusion gene or the NEG group.

Chapter 2

Affymetrix microarrays

The Affymetrix microarray technology platform is based on work done by Stephen Fodor and a team of scientists in the late 1980s. They realized that the fabrication processes used in the semi-conductor industry to create increasingly more transistors per surface could also be utilized in a biological setting. Advances in combinatorial chemistry helped them to put together a system that can simultaneously measure more and more different mRNA species in a biological sample. The measurement tools used by the technology are small oligonucleotide sequences (i.e., molecules composed of a chain of 25 nucleotides that are also referred to as "probes"). These oligonucleotides are made by a procedure which integrates solid-phase photo-chemical synthesis with photo-lithographic techniques.

A wafer of approximately 5 square inches is coated with a photo-sensitive substance. This substance in turn protects against a chemical coupling reaction when the compound ("protecting group") is intact on the surface. The synthesis of the sequences is controlled in a similar way in which computer microchips are made: masks with tiny gaps direct a light beam to defined positions on the chip surface. This selective exposure pattern determines the areas where the photo-sensitive substance is destroyed. In turn the area gets activated for a nucleotide coupling reaction. By step-wise covering the surface with one of the four nucleotides, the oligonucleotide chain is chemically built base by base as the newly coupled base again carries a photo-sensitive protecting group. For each subsequent synthesis step, a different mask is used and the complete 25 base nucleotides are built on the surface of the waver.

Such a photo-lithographic synthesis of oligonucleotides inherently carries the same potential as the photo-lithographic production of computer chips: with the establishment of the next technology level, the size of the area where the sequence is synthesized becomes smaller while exponentially increasing the number of probes that can be made on the array. While the first Affymetrix arrays had a feature size of 100 microns, the smallest feature size of current microarrays is 5 micron. Currently available commercial arrays carry over 6 million unique sequences that are synthesized on an array of about 1.7 cm^2 in size.

FIGURE 2.1: Image of a standard Affymetrix microarray. On a surface of 1.3 cm x 1.3 cm millions of probes with different sequences are present to quantify the abundance of all mRNA species incoded by the human genome. Image courtesy of Affymetrix.

2.1   Probes

In contrast to microarray technologies that use single, long probes for the detection of a given transcript, one characteristic feature of current Affymetrix microarrays is the measurement of a given transcript via a set of multiple 25-base long oligonucleotides ("probes") that vary in their sequence composition. One reason for choosing smaller oligonucleotides is the limited efficiency of the chemical coupling reaction by which the oligos are built step by step on the surface of the microarray. However, small probes also offer an advantage over long oligonucleotides as they are typically better in discriminating between similar, closely related transcripts, especially when dealing with mRNAs that are highly abundant in the sample.

For each mRNA species, a set of probes (probeset) is designed. It typically consists of 11 different probes that will specifically bind a single transcript at different positions along the mRNA sequence. Affymetrix microarrays usually contain tens of thousands of different probesets.

Probes are selected in such a way that a given sequence representing a portion of a gene sequence is completely complementary to the mRNA species that it is supposed to detect. These probes are called "perfect match" (PM) probes. Ideally a PM probe should be unique, thereby eliminating false signals from transcripts that could also bind to the probe due to complete sequence similarity. Especially the standard whole genome microarrays from Affymetrix still use another type of probes called "mismatch" (MM) probes. These probes are intended to detect the amount of unspecific signal and background signal detected by a PM probe. Subsequently one could subtract the MM intensity from the corresponding PM intensity. MM probes are also 25-base long oligonucleotides with the same sequence as the PM probe, but they contain a single modified base in the middle 13th position. Together a PM probe with its corresponding MM probe are referred to as a probe pair.

In the past years it has been shown, however, that the MM probes do not only detect unspecific signals but also detect the same transcripts as their PM counterparts[9]. Techniques that correct for unspecific signals by subtracting the intensity of the MM probe from the PM probe are faced with two problems:

1. The resulting measurement can be smaller than it actually is.

2. The variability of the measurement is doubled by combining the variable PM measurement with the variable MM measurement.

This observation motivated the development of many alternative approaches that utilize the PM measurements only. For more details on these techniques, please refer to Section 4.1.5.

When comparing the performance of various systems, it has become apparent that the sequence of the probe is actually the main cause for varying results between platforms. Besides the differences in sequences used by different systems to detect a transcript, the technologies also vary with respect to their ability to detect low expressed genes[15].

Probes vary in their ability to hybridize to the corresponding transcript (sensitivity). Factors such as the length of the probe, the extent of complementarity and the overall base composition play a role in capturing a target mRNA. Figure 2.2 shows for example the impact of GC content on signal intensity; the measured signal intensities clearly increase with increasing GC content. The plot is made using R code published in [7].

But also sequence-independent factors, such as the concentration of the probe on the array and the concentration of the target in the hybridization mix, contribute to the performance of the probe. Time, temperature, cation concentration, valency and character, pH, dielectric and chaotropic media, surface characteristics of the solid support, and density spacing of the probe molecules synthesized on the surface are many other factors that together determine how well a probe will capture its target. Furthermore, a sample-dependent complex background signal is caused by interactions of the probes with all the complementary sequences present in the sample. This background signal reduces specificity and sensitivity.

Figure 2.3 gives an example of how the different probes of a probeset measuring the expression for the gene transthyretin (Ttr) in dataset HD can differ in their performance. Furthermore, due to sequence similarities with other transcripts, there will be differences in the probes potential to bind transcripts that should not bind (specificity). For another example see also Figure 3.1.

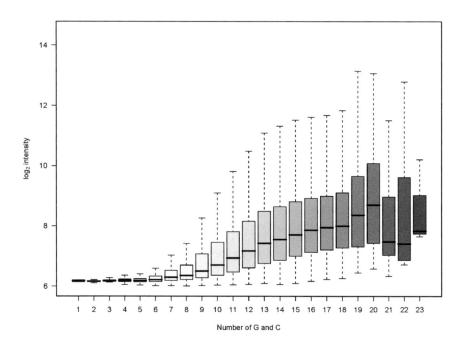

FIGURE 2.2: Impact of GC content on signal intensity. The more a given probe contains guanine or cytosine bases, the stronger the measured signal. The samples from the ALL dataset are used to construct for this graph.

All probes of a given probeset are distributed across the microarray and are not located next to each other. Thus, if there are issues with the data from an individual microarray such as wrong intensity measurements due to dust particles, background patterns or bubbles, the final number for the whole probeset will be less affected by the artifact.

Another source of variability comes from splice variants (see BioBox 2.1). If a cell produces, e.g., two different transcripts from the same gene which differ in how the exonic regions are spliced together, probes that will measure only one of the two splice forms will result in a weaker signal than probes that will measure both transcript forms.

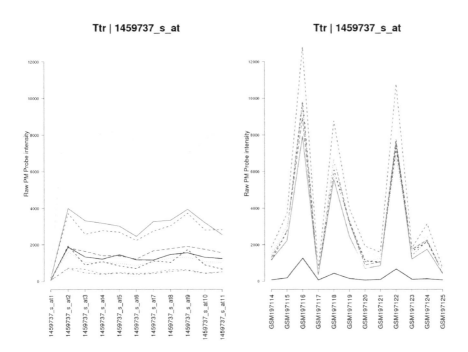

FIGURE 2.3: Differences in sensitivity of probes. Both graphs show raw signal intensities measured for the 11 probes of one probeset for the gene Ttr. The left graph shows the intensities for the different probes whereby the lines represent the 12 different samples of the experiment. The right graph shows the intensities for the different samples. The lines represent the 11 different probes of the probeset. The left graph shows clearly that probes show differences in how much signal they generate. The right graph highlights samples GSM197116, GSM197118 and GSM197122 as samples which show much higher expression for this gene than do the other samples. Ttr is a gene that is linked to a dissection artifact whereby high levels of the gene indicate that more cells from the surrounding tissue were dissected out than for the other samples in the experiment.

2.2   Probesets

Using multiple probes per transcript leads to multiple measurements which contribute to a more robust summarized intensity for the transcript and at the same time remove technical artifacts from the signal such as background. See also Section 4.1.5 on summarization. Among others, these reasons have contributed to the array design decision to have multiple probes for measuring a given transcript. The so-called probesets currently comprise mostly 11 different probes (for the current standard arrays) or 25 different probes (for the current Gene 1.0 ST arrays).

Even though the quality of the transcript measurement is enhanced by having multiple probes, the scientist still faces major questions when interpreting the data: How do different transcript splice variants affect the signal? How about the rare (but existing) problem of overlapping sense and anti-sense transcripts? How about changes in mRNA annotation for both adjustments of what is considered to be the correct sequence and the function of the encoded protein?

While it seems to be an impossible task to consider all of these issues for all of the measurements, one approach is to neglect such issues initially, but consider them after all data analysis steps have pinpointed the alterations in the experiment to the relevantly changing transcript. Another approach is to consider them from the start and apply alternative mappings of probes to probesets (see below in Section 2.2.2).

2.2.1   Standard probeset definitions

Scientists involved in the use of microarray technology and in the analysis of data generated by gene expression experiments need to decide on a strategy of how to deal with the ongoing changes in annotation. When using the links between the Affymetrix probeset identifiers and the transcripts that Affymetrix assigns to these probesets, it has been shown that, e.g., 5% have been changing over time when looking at the mouse genome 430 2.0 array[16]. Moreover, for the biological interpretation of the obtained results it is important to be aware of the fact that some probesets will detect two or multiple different transcripts as their probes are not specific. This is especially relevant when one uses the Affymetrix probeset definitions and looks at genes that have multiple probesets assigned to the same gene. Here one regularly is faced with the problem that occasionally probesets will disagree among each other while they ought to measure the same transcript based on their assignment to the same gene. Possible causes for such inconsistencies are unspecific probesets or probesets that detect different splice variants (see BioBox 2.1).

A starting point for a more detailed analysis of different probeset measurements for the same gene is looking at the actual name of the probeset. Table

Alternative RNA splicing. This term refers to the process by which a single gene can lead to several proteins. It is one of the processes resulting in the larger diversity of proteins in higher eukaryotes. It takes place prior to the translation of the mRNA sequence into protein. The first transcript of a gene is called a "pre-messenger RNA (pre-mRNA)." At this moment, it includes still sequences that are not coding for the protein. These non-coding pieces are called "introns," while the coding portions of the transcript are called "exons." In an event referred to as RNA splicing, the non-protein-coding pieces of sequence are removed. However, the process, by which the protein-coding pieces are joined, gives rise to a new form of diversity, called "alternative RNA splicing." In essence, it is possible that not all exons end up in the final transcript and/or that the order of the sequence portions encoded by the individual exons can be rearranged. In other words, a single pre-mRNA can lead to a diverse set of different mRNAs (called "splice variants[a]") which in turn produce proteins with at times different biological functions[17].

A well-known example of alternative splicing is the gene Bcl-x. The encoded regulatory protein can have one of two functions, depending on what isoform is expressed. While Bcl-xL produces a larger isoform which suppresses programmed cell death (apoptosis), the smaller isoform Bcl-xS will result in a regulatory protein which actually promotes apoptosis.

---

[a]Alternative splicing can lead to impressive diversity: The *Drosophila* gene Dscam can produce approximately 38,000 different, alternatively spliced transcripts.

BioBox 2.1: Splice variants

2.1 gives an overview of the different types of probes sets, the name by which they can be recognized and what their characteristics are for the currently used designs. Table 2.2 gives an overview of the old probeset names that are now retired.

A second step is to consult the most up-to-date annotation of Affymetrix as they provide each probeset with an annotation grade and much further detail via their Internet site NetAffx[18] (see Table 2.3).

2.2.2   Alternative CDFs

As mentioned in the previous section, already during the chip design process it was unavoidable to postulate sets of probes that were less specific than others. Some of these probesets were unspecific due to a lack of knowledge about the accurate position of the corresponding probes on the transcript or the matched genome position. Another major cause was the concept of the arrays: The "whole genome" arrays were intended to do just that: attempt to cover the whole genome. This is impossible, e.g., for gene families, where a number of genes are highly similar to one another when looking at their gene sequences. Accordingly, it was impossible to design sets of probes that were specific for each member of the gene family.

What is a solution to the problem when one intends to make an array layout that claims to cover the whole genome? The only possible solution is to design probesets that attempt to be as specific as possible, but drop the specificity rules for members of closely related gene families. Affymetrix has chosen this approach for their standard arrays. Of course, the specificity of the probesets are indicated both by the name of the probeset as well as the annotation that is quarterly updated. For some time the problem of unspecific probes has become worse with every revision of the genome annotations whereby gene positions have been adjusted while the chip layout remained unchanged. Typical problems that have arisen are probes that are now located outside of the region that codes for a mRNA, probes that are detecting the wrong strand of the appropriate DNA sequence, probes that are unspecific for a single transcript, etc. However, as the annotation of the human genome stabilizes, an increase in the detection of incorrect probes will become ever smaller.

For researchers who do not want to choose a solution which attempts to cover every gene in the genome, but rather want to opt for less, but more accurate information, a solution is to redefine the mapping of probes to a probes set. Such newly defined probesets contain only those probes that are known to map to unique sequences in the genome or can be mapped to transcripts or genes based on one of the public databases. Depending on the purpose of the microarray experiment (i.e., the question that is to be answered by doing the biological experiment and running the arrays), one should consider using those alternative mappings by using so-called "alternative CDFs" (or "custom CDF"). Dai et al.[19] have come up with such a different assignment of probes to new probesets based on various different databases, such as En-

TABLE 2.1: Currently used Affymetrix probeset types and names. The first column refers to the name endings that denote a certain type of probeset. The second column gives information on how such a probeset was designed. The third column highlights potential difficulties in interpreting the signal obtained by such a probeset. The last column refers to the arrays for which these probeset types are used.

Name suffix	Description	Issues	Arrays
_at	Unique probeset. These probesets contain probes that specifically measure unique sequences of a single target transcript.		HG-U133
_a_at	"Alternative transcripts" probeset recognize alternative transcripts for the same gene. These probesets are a subset of the _s_at probesets of the non-"Plus" arrays. Probesets of the _s_at probeset type that were already present on the HG-U133A and B arrays were not changed to _a_at probesets for consistency reasons.	The signal cannot be attributed to a single transcript and is therefore difficult to interpret.	"Plus" arrays
_s_at	"Shared" probeset against two or more transcript variants sharing a common sequence. Probes in these probesets are therefore common among multiple transcripts that can originate from different genes.[1]	The signal cannot be attributed to a single transcript and is therefore difficult to interpret.	HG-U133
_x_at	Probeset containing probes that are identical or similar to highly unrelated sequences. Rules for cross-hybridization were dropped.	Besides the target transcript, various other transcripts may hybridize to the probes in an unpredictable manner resulting in a signal that can be difficult to interpret.	HG-U133

[1] Approximately 90% represent splice variants and the rest interrogate alternatively polyadenylated transcripts or highly similar transcripts.

TABLE 2.2:   Affymetrix probeset types and names that are not used in current designs. These names were used in designs like the U95 arrays. The first column refers to the name endings that denote a certain type of probeset. The second column gives information on how such a probeset was designed. The last column highlights potential difficulties in interpreting a signal obtained by such a probeset.

Name suffix	Description	Issues
_b_at	"Ambiguous" probeset containing probes for which the probe selection rules had been dropped.	The signal cannot be attributed to a single transcript and is therefore difficult to interpret.
_f_at	"Sequence family" probeset containing some probes that are similar, but not necessarily identical to other gene sequences.	The signal cannot be attributed to a single transcript and is therefore difficult to interpret.
_g_at	"Common groups" probeset containing probes chosen in a region of overlap between transcript sequences. These probesets will always have a unique _at probeset on the array as well.	The signal cannot be attributed to a single transcript and is therefore difficult to interpret. It will typically be more informative to look at the corresponding _at probesets.
_i_at	"Incomplete" probeset containing fewer than the required number of unique probes.	The signal cannot be attributed to a single transcript and is therefore difficult to interpret.
_l_at	"Long" probeset where a sequence was represented by more than 20 probe pairs.	
_r_at	"Rules dropped" probeset where it was not possible to select a probeset without dropping some of Affymetrix probe selection rules.	The signal cannot be attributed to a single transcript and is therefore difficult to interpret.
_s_at	"Shared" probeset designed against two or more transcript variants sharing a common sequence. This probeset type was the equivalent of the current _a_at probeset type. Probes in these probesets all matched multiple transcripts from the same gene, gene family or homologous genes.	The signal cannot be attributed to a single transcript and is therefore difficult to interpret.

TABLE 2.3: Annotation grade of original Affymetrix probesets as provided by NetAffx. There are five levels of relationships between probesets and the current transcript record. For each probeset only the transcripts with the highest available assignment grade are referenced in NetAffx.

Grade	Description
A	Matching probe: Nine or more probes from the probeset match this transcript perfectly.
B	Target-transcript overlap: The transcript and the target sequence of the probeset overlap on the genome. The probes do not match the transcript, presumably because the 3' end of the transcript is truncated in the record.
C	Consensus-transcript overlap: The transcript and the consensus/exemplar sequence of the probeset overlap on the genome. The transcript does not overlap the target sequence presumably due to 3' end of the truncation.
E	No transcripts are known to correspond to this probeset at this time, but a UniGene EST cluster is known to correspond to it.
R	No transcript currently supports this probeset, though EST sequences are available from the design information.

sembl Transcript or Entrez Gene. Instead of representing a maximum number of transcripts using the general rule of 11 PM and 11 MM probes per probeset as Affymetrix has done in their chip design, these mappings are based on the most current genome annotations and, among many other requirements, omit all probes that are not unique for a given transcript. Table 2.4 gives a summary of the rules that are used to generate these alternative mappings.

The most recent version of the mappings done by Dai et al. includes also probes containing one-mismatch probes outside of the central 15 bp region. This change is based on thermodynamics analyses suggesting that probes with mutations in the periphery of the probe do not significantly reduce hybridization signals. The authors expect to see that probes with one non-central region mismatch will reduce the influence of cross-hybridization on the final results. And, of course, using such one-mismatch probes will also increase the total number of probes that can be utilized[20].

When looking at the performance of these alternative probeset definitions, one can see a clear improvement in the accuracy of detecting differential gene expression as well as the variability and reproducibility of the obtained gene expression measurements[21]. Using, e.g., the Entrez Gene-based probe mappings, one chooses to ignore any differential signal due to splice variants on one hand while obtaining a single value for a given gene on the other hand. This characteristic is especially useful when making use of pathway analysis algorithms and programs as they tend to expect a single measurement per gene and are usually not capable of dealing with splice variant data.

TABLE 2.4:   Rules for generating alternative CDFs. These rules are used by Dai et al.[19] to generate CDF files based on UniGene (most similar to the Affymetrix annotation) or, e.g., Entrez Gene. The rules given here as an example for Entrez Gene are also used for many other alternative mappings such as Ensembl Gene, VEGA Gene, etc.

CDF	Rule
UniGene	A probe must match perfectly to both cDNA/EST[2] sequences and genome sequence.
UniGene	A probe must only hit one UniGene cluster and one genomic location.
UniGene	All probes representing the same gene must align sequentially in the same direction within the same genomic region.
UniGene	Each probeset must contain at least three oligonucleotide probes. Probes in a probeset are ordered according to their genomic location.
Entrez Gene	A probe must hit only one genomic location.
Entrez Gene	Probes that can be mapped to the same target sequence in the correct direction are grouped together in the same probeset.
Entrez Gene	Each probeset must contain at least three oligonucleotide probes. Probes in a probeset are ordered according to their location in the corresponding exon.
All	Probes that have only one mismatch at location 1-5 or 21-25 are included.

[2]EST = Expressed Sequence Tag.

It still needs to be investigated further to what extent there is an impact of the variable probeset sizes for the different summarization techniques as the alternative CDFs contain probesets that range from a minimum of three probes to tens of probes (see Figure 2.4). For sure one can expect to see an impact on the standard error of the signal estimate depending on the probeset size.

The probesets interrogate various parts of transcripts and we know that there are different ways of how the cell alternatively assembles the coding transcripts (i.e., alternatively spliced transcript variants). There is no such thing as uniform gene-level expression. Perhaps in some cases just a specific alternatively spliced transcript was significantly expressed.

With the increasing number of probesets, the utilization of probes has remained fairly stable at around 45% of all the probes on the array (see Table 2.5). The major reason for this drastic reduction in probes is due to the removal of all the mismatch probes as well as all unspecific probes. Still, the quality of the biological information has improved. Especially when using gene-level alternative CDFs such as the version based on Entrez Gene or

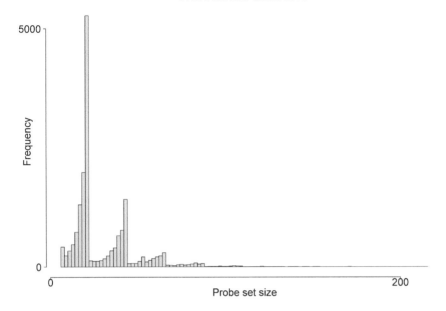

FIGURE 2.4: Differences in probeset sizes when using alternative CDFs.

Ensembl Gene, it is much more straightforward to use downstream pathway analysis tools, as there is only one measurement per gene. Furthermore, false negative results are avoided that are generated by Affymetrix annotation linking a given probeset to a gene which contains a number of non-specific probes.

2.3 Array types

Affymetrix has developed a number of different microarrays for gene expression studies. They differ in quite a number of aspects such as the number of different probe sequences per array, the use of mismatch probes vs. background probes, the focus on genes, exons or genome-wide transcription, the laboratory processes for preparing the hybridization mixture, etc.

Standard Expression Monitoring Arrays The majority of full-length and partial gene sequences contained in publicly available sequence databases, which correspond with human, mouse, rat, canine, *Drosophila melanogaster* (fruit fly), *Xenopus laevis* (frog), *Danio re-*

TABLE 2.5:   Probe utilization for human array alternative CDF based on Ensembl Gene. Values for Ensembl Gene-based probe utilization are based on the probe count for the CORE division of Ensembl as the target numbers are also only derived from the CORE division of Ensembl. The percentage is relative to the total number of 604,258 probes on the array. The data for versions 3, 7 and 8 were omitted as they used different metrics.

Version	Probesets	Probe utilization (%)
1	16,698	45
2	16,308	42
4	16,725	42
5	16,451	40
6	16,609	40
9	17,209	46
10	17,215	46
11	17,429	46

rio (zebrafish), *Saccharomyces cerevisiae* (yeast), *Escherichia coli* (bacteria), *Pseudomonas aeruginosa* (bacteria), *Plasmodium falciparum* (malarial parasite), Anopheles (mosquito vector of malaria) and *Arabidopsis thaliana* (plant) organisms are covered by these types of arrays.

Custom Express Arrays These focused arrays are designed for a specific purpose by the researcher and are produced by Affymetrix. They typically include only a subset of genes and tend to be less expensive when ordered in a large batch as they can be realized using smaller chip sizes.

Exon Arrays Using the smallest currently used feature size of 5 micron, these arrays are intended to provide gene expression information on the exon level. Global expression of individual exons can be monitored to identify alternative splicing events and to obtain genome-wide exon-level expression data that in combination with sequence information has the potential to reveal the phenotypic splicing patterns resulting from genetic variations.

Gene Arrays In comparison to the standard arrays, the gene arrays have more accurate coverage of the transcribed genome as they are based on most recent genome annotations for mouse, rat and human. Since they do not contain mismatch probes while using a 5-micron feature size, the size of the chip is smaller than the standard arrays. Besides a smaller prize, they also require a smaller hybridization volume. In contrast to the IVT procedure that incorporates a 3' bias in the obtained labeled probe, the Whole Transcript (WT) Sense Target Labeling Assay produces a more complete and accurate view of a gene's total transcriptional activity.

Tiling Arrays These arrays are available for a few organisms including human, mouse, drosophila, *C. elegans* and yeast. Instead of positioning the probes according to information about genes and expressed sequences, these array designs use evenly spaced probes across the non-repetitive portion of the genome. They rely purely on genomic DNA sequence and not functional annotation for their design. Typical applications include mapping of the entire collection of transcribed elements (including noncoding RNAs), identifying protein binding and methylation sites and identification of genomic origins of replication.

2.3.1 Standard expression monitoring arrays

The probes for the standard arrays (e.g., for mouse, human or rat) were designed using public resources such as UniGene[22], GenBank®[3][23], dbEST[24], and RefSeq[25]. In this process sequences (e.g., cDNA sequences) from the databases were aligned to form clusters. These clusters were matched to genes, after low-quality regions of EST sequences were removed and orientation were checked. As the quality of the sequences included in a cluster varied (the number of mismatches with other sequences of the cluster or the length of a given sequence), the final probe selection was based on a consensus sequence for the cluster, also referred to as exemplar sequence. The lab procedure for running these arrays makes use of the IVT kit, which introduces a bias towards the 3'-end. This characteristic was taken into consideration during the probe selection process by selecting probes that are most proximal to the 3'-end of the transcript.

When one does not specifically need the information about a splice variant or when one is not interested in detecting novel transcripts, the standard/traditional whole genome arrays are a good compromise between ease of analysis due to smaller data amounts and interpretability of the findings based on the fact that most scientific knowledge is currently based on gene level information available through databases such as Entrez Gene.

2.3.2 Exon arrays

The working copy of a gene, the mRNA, is used to synthesize a protein. Prior to using this working copy, a processing step is carried out in the cell which removes non-coding sections of the mRNA. This process is called "splicing" and is done for about half the genes of the human genome. The sections that are excised are referred to as introns while the exons remain in the mRNA and contain the information for the construction of the protein. Alternative splicing of the exons of a single gene can result in different mRNA species

---

[3]Registered trademark of the U.S. Department of Health and Human Services, Bethesda, Maryland.

that will therefore encode different final proteins. It has been shown that alternative splicing takes place in at least 74% of human genes with multiple exons[26] and more recent publications predict up to 94% of all human genes to generate more than one product[27].

Exon arrays can be used to study the role and the regulation of different splice forms and their prevalence in different tissues and during different stages of developmental processes. They can be a useful tool in studying alternative promoter usage and alternative termination of transcription. Besides containing probesets based on data from RefSeq[25], the exon array also contains probesets for exons predicted by algorithms such as Genscan[28], Twinscan[29], and GeneID[30][4]. It is important to realize that these gene prediction tools are not perfect and will therefore result in a number of probes and probesets on the Exon array that will not detect a transcript. This together with the Affymetrix attempt to represent every exon on these arrays, a lot of probes have characteristics that are not optimal for the system resulting in more than 80% of probes that are either below the detection limit or show intensities that are at the level of background[31].

Instead of having one or multiple probesets for a gene as with the standard arrays, the exon arrays aim to provide the researcher with gene expression information at the exon level. In other words, Affymetrix focused on exons for the design of the exon arrays while the gene and the known transcripts were at the center of the design of the standard whole genome arrays. In consequence, the use of exon arrays is advisable when one is interested in obtaining insight in differences between sample groups that are likely to be linked to different isoforms or splice variants of a gene.

While the standard arrays are run with a hybridization cocktail containing anti-sense nucleic acids complimentary to the probes on the array, the Gene 1.0 ST and the Exon 1.0 ST arrays are used with a laboratory protocol that produces sense targets (see also BioBox 2.2).

Instead of having mismatch probes to estimate the amount of unspecific hybridization, both the Gene 1.0 ST and the Exon 1.0 ST arrays have a number of probes with various GC contents that are either mismatch probes against a few transcripts or probes that carry a sequence that is not complementary to a variety of different genomes including human, mouse and rat. These probes are referred to as background probe collection (BGP). One major advantage of replacing MM probes with this alternative set of background probes is the

---

[4]These computational gene recognition software tools are used to predict the location of a protein-coding regions in uncharacterized DNA sequences. They either work only with DNA sequence information by making use of sequence patterns or they utilize the results of similarity searches with sequences of open reading frames (the sequences that are flanked by a START codon and a STOP codon). In contrast to prokaryotic organisms, gene prediction is more difficult in eukaryotic organisms due to a much lower gene density. Furthermore, short exons are difficult to identify.

Some gene characteristics. Genes contain the information, e.g., for the synthesis of proteins. This unit of inheritable information is encoded in the molecule DNA. There are typically two types of information that belong to a given gene: "coding" sequences contain the instructions how to build, e.g., a protein while "non-coding" areas play a role in the regulation of the activity of the gene. As many molecular processes (e.g., DNA replication) in a cell occur only in one direction, molecular biologists refer to the different ends of a stretch of DNA according to what chemical group is present at its end: the 3'-end exhibits a hydroxyl group while the 5'-end exposes a phosphate group. When the sequence of a gene is transcribed into a mRNA molecule, its sequence of nucleotides is called "sense" as it is the template for making the gene product (protein). In other words, a nucleotide sequence that is translatable into protein makes "sense." Its complementary sequence is called "antisense." Therefore the DNA sequence that serves as a template to make a transcript is actually containing an antisense sequence as the complementary sequence becomes the sense mRNA sequence.

The combination of genetic information with environmental factors is thought to be responsible for how an organism develops and displays different characteristics (phenotype).

BioBox 2.2: Gene

substantially smaller amount of probes on the array. This allowed almost a doubling of PM probes that could be put onto the array.

Since the exon arrays were designed based on a more recent and therefore more accurate annotation of the human, mouse and rat genomes, another alternative for using the arrays other than obtaining exon level information is to get more accurate gene-based transcription data. As the exon arrays carry many more probes per gene, one can be much more stringent in requiring specificity for the probes that comprise gene-level probesets. Furthermore the probeset size increases clearly resulting in much more robust measurements per probeset/gene.

Exon arrays provide the researcher with a tool that assesses the transcriptional activity of an entire genome at the level of individual isoforms and exons. They cover the whole gene locus while the standard whole genome arrays have a selection bias of probes towards the 3'-end of a gene. Furthermore, exon arrays do not contain paired mismatch probes. Due to this change, there

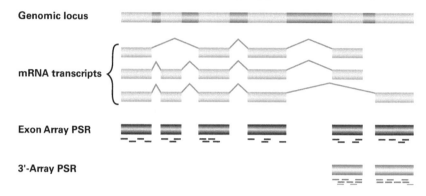

FIGURE 2.5: Schematic for coverage of probesets across the entire length of the transcript for exon arrays and 3' arrays. Golden regions are exons, whereas the grey regions represent introns that are removed during splicing. The short dashes underneath the exon regions for the Exon Array and 3'-Array probe selection region indicate individual probes representing that region. Image courtesy of Affymetrix.

are no absent/present calls available. This has been replaced by a different method which looks at the signal above background (DABG) to come up with an indication about the reliability of a given measurement (see Section 4.1.7). While the number of oligos per probeset has been reduced from 11 (as for the whole human genome array HGU133plus2) to 4, the total coverage per gene has gone up to an average of 40 probes (see Figure 2.5). For example, the Human Exon 1.0 ST array contains 1.4 million probesets for known and predicted human exons.

In addition to the normal mapping of probes to probesets, Affymetrix has arranged probesets into "meta probesets" with different levels of complexity. This is again linked to the design difference with the standard arrays: as the design is exon focused, there are no transcript or gene entities. Rather exon-specific probesets are grouped into "meta probesets" afterwards. While the probesets are fixed the meta probesets evolve with every new genome assembly or genome annotation update. As there are different levels of confidence attached to the different annotation sources, Affymetrix has chosen to come up with a hierarchy of gene confidence levels. This is important to understand as the inclusion of probesets with little confidence can have a negative impact on the summarized value of the meta probeset. The "core" set of meta probesets includes exons of well-characterized transcripts (RefSeq and full-length mRNAs). The "extended" set includes alternatively spliced mRNAs and all other cDNA-based annotations. The "full" set finally contains all exons independent of their biological relevance. The full set also contains exons that were defined purely based on exon prediction algorithms and at times even suboptimal predictions. For each region of a given gene the derived annota-

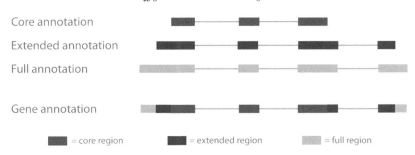

FIGURE 2.6: Transcript annotations derived from different confidence levels are merged to form a gene annotation. Each gene region is labeled according to the highest confidence level for that region. Image courtesy of Affymetrix.

tion is labeled according to the highest level of confidence that was available for that region (see Figure 2.6).

These different types of meta probesets are important to understand when choosing to generate gene-level expression intensities (as opposed to exon-level expression intensities) with the tool ExACT from Affymetrix (instead of other tools such as Bioconductor/R). Conceptually these gene-level estimates generate something similar to the alternative CDF mappings discussed in Section 2.2.2 for standard IVT-based arrays: the researcher obtains single expression measurements covering the entire gene region (instead of only the region covered by the primarily 3'-end biased probes on the standard arrays). It is important to note that there is currently not a single best way to come up with gene level estimates. A good start is certainly the use of the core set of meta probesets or the use of alternative CDFs based on, e.g., Entrez Gene information. This will help to establish the correct execution of the experiment. Once one is confident of the findings of the study, one can continue looking into the meta probesets or individual probesets of more exploratory nature.

However, when interpreting the findings of Exon microarray experiments, it is currently still difficult to put findings, e.g., related to potential alternative splicing events into context of biological knowledge such as known pathways since our understanding of the function of different splice variants of the same gene is still much less than our limited knowledge of a gene without considering its splice variants. Still, even when these shortcomings are dealt with, the majority of the topics discussed in this book will remain applicable to Exon arrays as well.

2.3.3 Gene arrays

For the next generation of standard whole genome arrays (e.g., the Human Gene 1.0 ST) Affymetrix has dropped the mismatch probes in favor of a larger number of perfect match probes for every single gene. Instead of having 11 probes within a probeset that were selected to be targeting the 3'-end of a transcript, one now obtains data from approximately 26 different probes covering the whole transcript length (see also Section 3.2.2 on relevant details about the protocol). Removing the 3'-bias should result in more accurate gene level data especially in cases such as alternative splicing at the 3'-end of the gene, alternative polyadenylation sites or non-polyadenylated transcripts, etc. From a practical point of view it should also provide the researcher with more robust measurements from samples with degraded RNA.

Approximately 80% of probes were selected based on probes that are also present on the Exon arrays. Therefore the laboratory process for labeling the RNA is also making use of the whole transcript labeling kit (WT) as opposed to the *in-vitro* transcription kit (IVT) which is most suitable for probes located close to the 3'-end of a gene. The layout of these arrays is based on more recent versions of genome sequence assemblies as compared to the standard arrays.

Besides interrogating the expression level of 28,869 well-annotated genes, Affymetrix has also included some 20,000 generic background probes for background estimation on these arrays. These probes are selected in a way that they are not present in any of seven major genomes (human, mouse, rat, *Drosophila melanogaster*, *Caenorhabditis elegans*, *Saccharomyces cerevisiae*, *Arabidopsis thaliana* and *Escherichia coli* K12) and also are unlikely to cross-hybridize with transcripts of any of these genomes. Twenty-six bins of approximately 1,000 probes with varying GC content (between no G or C nucleotides and only G or C nucleotides) for each bin were chosen.

The small variance seen with the HG-U133 Plus 2.0 arrays in comparison to the Human Gene 1.0 ST arrays is most likely due to the larger feature size: instead of 11 micron by 11 micron features the size is now 5 micron by 5 micron. However, since the Human Gene 1.0 ST arrays have many more probes per gene, the variance on the gene level is similar[31]. A benefit of the smaller feature size is that the whole array format is smaller than the standard HG-U133 Plus 2.0 arrays and therefore typically also somewhat less expensive. In contrast to the 1,000 ng of total RNA of starting material and an obligatory rRNA reduction step for Exon arrays, the protocol for Gene arrays requires only 100 ng of total RNA as starting material and a step for reducing the amount of ribosomal RNA is not needed. This is advantageous as it not only requires less RNA, but also omits an extra processing step that can be a source of non-biological variability.

2.3.4   Tiling arrays

While exon arrays and standard gene expression arrays focus on studying the activity of genes that encode for proteins, it has become clear over the last years that much more genome-wide transcription takes place. Studies in various species have shown that, e.g., the human genome is transcribed on both strands generating transcripts that have little or no protein-coding potential. Besides discovering novel RNA species, it has also been shown that there are sequences that serve multiple purposes. Both strands can encode for transcribed sequences and regulatory elements and these stretches can overlap as well[6],[32].

Tiling arrays make use of 25-mer probes that are evenly spaced every 35 bases along the complete genome with a gap of approximately 10 bases between each probe. To avoid uninformative signal from repetitive sequences, a program was utilized that removed such elements from the array design. Furthermore, probes were avoided that would likely show non-linear increases in signal intensity with increases in transcript concentration. Due to the enormous amount of different probes, the coverage of whole genomes such as the human tiling arrays needed to be split up into several arrays.

Among other applications, novel types of RNA species (e.g., non-coding, functional RNAs) can be studied with tiling arrays that were made available in 2005 by Affymetrix together with the exon microarrays. While exon arrays offer the potential to detect differences in abundance of splice variants of the same gene, tiling arrays can be used to detect novel transcripts encoded by genes that do not contain the information for proteins or enzymes. They can also be used to search for transcription factor binding sites, DNA methylation profiling or used in assays that identify transcripts that are bound by RNA binding proteins.

Even though both applications are scientifically very interesting, the general use of these microarrays is currently hindered by a few issues which need to be considered, when choosing to use either tiling or exon arrays, such as large data amounts, expensive consumables, fewer data analysis tools, or sequence duplications that make it difficult to assign a given probe signal to a single genomic location.

2.3.5 Focused arrays

Focused arrays (also referred to as "theme chips," "diagnostic microarrays" or "boutique arrays") typically contain only probesets against a small selection of well-annotated genes. These can be for example relevant for a particular pathway (e.g., cholesterol synthesis), contain all members of a gene family (e.g., kinases), represent a specific function (e.g., toxicity genes) or comprise candidate genes that were obtained using whole genome screening experiments.

They are attractive from a data analytical point-of-view as the number of genes relevant for a certain biological experiment is highly enriched, which in turn reduces the number of false positive findings and the extent of correction for multiple testing (see Section 5.5.3). For the biologist the focused arrays are attractive as they usually are smaller arrays and therefore tend to be much less expensive. The smaller size of the array also has practical implications in the lab as they require less starting material (i.e., total RNA), are scanned faster and therefore tend to allow for higher throughput.

On the downside, it has to be noted that due to the selection of genes for a particular topic, it usually occurs that more than half of the genes show differential expression between sample groups. Furthermore, it also is not uncommon that the observed gene regulation is predominately going into one direction. Therefore, focused arrays require a different normalization strategy such as using the intensities obtained from probesets detecting transcripts that were spiked into the samples right after RNA isolation (see also Section 4.1.4).

2.4   Standard lab processes

The following section gives an overview of the standard steps that are involved in running gene expression profiling experiments on the Affymetrix platform. As the standard IVT process has been accompanied by a new whole transcript procedure, we will cover the common steps below and discuss the differences in subsequent sections.

After the biological experiment has been run and samples have been obtained, RNA is isolated. The RNA quality and quantity is checked for each sample and across all samples in an experiment. Good quality RNA is labeled according to one of the two procedures below.

Once labeled material is prepared a hybridization cocktail is mixed that contains the labeled target, salts, blocking agents and bacterial RNAs in known concentrations to ensure the quality of the subsequent steps. This mixture is injected into microarray cartridge. It is then hybridized at 45°C for 18 hours. After hybridization, the chip is stained with a fluorescent dye (streptavidin-phycoerythrin) that binds to biotin which had been incorporated during the labeling process. The staining protocol includes a signal amplification step makes use of an anti-streptavidin antibody and a biotinylated IgG antibody. The fluorescent dye will then emit light when the chip is scanned with a confocal laser. The position and intensity of each signal on the array is recorded. This workflow from sample RNA to array data requires approximately 2.5 days working time.

2.4.1   In vitro transcription assay

The traditional protocol used for standard expression monitoring arrays involves the synthesis of double-stranded cDNA from the RNA sample using reverse transcriptase and an oligo-dT primer. This double stranded cDNA serves as a template in an *in vitro* transcription (IVT) reaction that produces amplified amounts of biotin-labeled antisense cRNA or aRNA (for amplified RNA) target. Prior to hybridization, the cRNA is fragmented using heat and magnesium ions. The fragmentation step reduces the cRNA to 25-200 bp fragments. Only fragments of the right size are able to efficiently and reproducibly hybridize to the probes on the array.

Affymetrix has recently released the second generation of this protocol. Referred to as "IVT Express Kit," the major advantages of this new protocol are the reduced amount of starting material (from 1 μg with the first generation kit down to 50 ng total RNA with the second generation protocol) and a reduction in time necessary to complete the protocol when using larger amounts of starting material. As the new kit allows for a wider range of amounts in starting material between 50 ng and 500 ng, it is necessary to use the same amounts for all samples in an experiment as the resulting data is somewhat influenced by the amount of starting material that is used.

2.4.2 Whole transcript sense target labeling assay

This assay is designed to avoid the 3'-bias of the IVT protocol. It also amplifies the starting material and generates biotinylated target DNA. However, sense-strand DNA targets are generated resulting in DNA/DNA duplex formation during hybridization in contrast to DNA/RNA heteroduplexes that are formed when hybridizing antisense-strand RNA as produced by the IVT protocol.

By using random primers instead of oligo-dT primers, targets are generated along the entire length of a transcript. As little as 100 ng of total RNA starting material is required for the assay. The protocol begins with the addition of poly-A RNA controls to total RNA and is optionally followed by a ribosomal RNA reduction step for the exon arrays. This step can be avoided when preparing target for the Gene arrays. Subsequently, cDNA is synthesized and IVT is performed. After a purification step, the antisense RNA is used for another round of cDNA synthesis. After fragmentation and labeling, the sense-strand cDNA target is hybridized to either exon or gene arrays.

2.5 Affymetrix data quality

2.5.1 Reproducibility

Various publications have looked into the reproducibility of the Affymetrix platform. Provided that the RNA used for the microarray experiment is of good quality, data from Affymetrix microarrays have consistently been shown to be highly reproducible (e.g., [33]).

One of the largest studies investigating the reproducibility of Affymetrix microarrays and other microarray platforms was the FDA-sponsored microarray quality control study (MAQC, see also Section 7.2). In this study two distinct reference RNA samples were analyzed. The main findings were high inter-platform concordance and high intra-platform reproducibility across different test sites using the same technology platform. Furthermore, even for discordant findings between lists of differentially expressed genes, the biological alterations that were identified were consistent despite platform dependent differences. Most remarkable for the Affymetrix platform were the high level of reproducibility and consistency within and across different test sites (figures 1 and 2 in [34]).

2.5.2   Robustness

A particular characteristic of the Affymetrix system is the redundancy in measuring the same transcript. For example, 11 perfect match probes and 11 mismatch probes are present on the human genome array HG-U133 Plus 2.0 for every transcript that is to be measured. The almost random distribution of these probes across the microarray surface in combination with the fact that all probes have a different sequence allows for a robust estimation of the abundance of the measured transcript. This robustness has been shown in the first MAQC study where no data of the data-generating sites were removed or repeated (tables 1 and S3 in [34]).

2.5.3   Sensitivity

Affymetrix aims to provide arrays that have a minimum sensitivity of 1:100,000. This concentration ratio corresponds roughly to a few copies of transcript per cell, or an approximate 1.5 pM concentration. Part of the data analysis of the MAQC study also looked at sensitivity (Section H in [35]). From this data it can be concluded that the platform is comparably sensitive in detecting changes in gene expression.

Chapter 3

Running the experiment

3.1 Biological experiment

3.1.1 Biological background

3.1.1.1 Aim/hypothesis

Classical molecular biology experiments were designed to quantify how much of a single type of molecule (e.g., a protein or a mRNA encoded by a gene) is present in a sample. Typical techniques for these types of experiments are Northern blotting or quantitative reverse transcription polymerase chain reaction (RTqPCR; see BioBoxes 3.1 and 3.8). The introduction of microarray technology brought about the possibility to measure very many different types of molecules of one chemical class in a single sample, e.g., all mRNA transcripts encoded in the genome. These new whole genome arrays can now be used to run experiments for hypothesis generation rather than hypothesis validation.

With microarrays it is possible to do a biological experiment and assess the consequences of a given stimulus on a sample without any prior assumptions as to what genes should be affected. On the other hand, the amount of data obtained in a typical microarray experiment is so large that it is impossible to look at all measured data points one by one. Therefore, it is necessary to be as clear as possible about the purpose of an experiment. Being able to tightly control all experimental variables and only alter the variable that is to be studied is essential to avoid misleading results. The majority of this chapter deals with various aspects that will ensure that the biological experiment is properly controlled.

3.1.1.2 Technology platform

Even though microarrays are perfect tools for hypothesis generation, they clearly have their limits. When studying the alteration of genes that are known to be expressed at low levels or if the alteration is expected to be limited (typically less than 50% increase or decrease in expression), microarrays can be the wrong tool. Microarrays have been shown to be less sensitive than quantitative RNA assays that measure only one or a few genes at a time[15].

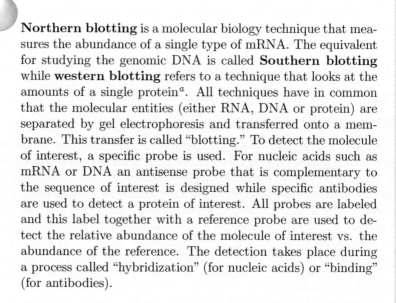

Northern blotting is a molecular biology technique that measures the abundance of a single type of mRNA. The equivalent for studying the genomic DNA is called **Southern blotting** while **western blotting** refers to a technique that looks at the amounts of a single protein[a]. All techniques have in common that the molecular entities (either RNA, DNA or protein) are separated by gel electrophoresis and transferred onto a membrane. This transfer is called "blotting." To detect the molecule of interest, a specific probe is used. For nucleic acids such as mRNA or DNA an antisense probe that is complementary to the sequence of interest is designed while specific antibodies are used to detect a protein of interest. All probes are labeled and this label together with a reference probe are used to detect the relative abundance of the molecule of interest vs. the abundance of the reference. The detection takes place during a process called "hybridization" (for nucleic acids) or "binding" (for antibodies).

[a]Southern blotting is a technique named after Edwin Southern who invented this technique in 1975. The other techniques have essentially been named in reference to Ed Southern with a sense of humor. J.C. Alwine developed a similar technique for RNA and called it not unintentionally "northern blotting." W. Neal Burnette then went on to call the protein blotting technique he developed "western blotting" as it was somewhat inspired by the northern blotting procedure. There is no widely known "eastern blotting" (some people propose to use this for a technique that uses antibodies to detect polysaccharides in a similar way as "western blotting" uses antibodies to detect proteins) but there is a southwestern blot for DNA-binding proteins.

BioBox 3.1: Northern blotting

Especially when one is interested in only a small number of low expressed genes, the use of RTqPCR is probably a better alternative.

Gene expression microarrays measure mRNA abundance. As there is only a limited correlation between mRNA abundance levels and protein abundance levels[36], scientists using microarrays for gene expression studies almost exclusively look at changes in mRNA levels between groups of samples. The assumption is made that changes in transcript copy numbers upon a stimulus/treatment reflect the intention of the cell to ultimately change the abundance of the protein encoded by a given mRNA. Therefore such changes should also be reflected in protein changes.

If, however, one is interested in quantitative measurements of transcripts, gene expression measurements from microarrays are likely to be inaccurate. There are however ongoing efforts that use physical modelling approaches instead of statistical techniques to use the signal intensities of Affymetrix arrays for quantification of mRNA abundance. Such an approach makes use of the physical properties that determine differences in signal intensities between different probes that measure the same transcript (in other words: probes belonging to the same probeset)[1]. For these quantitative models the physical process is mathematically formulated. This includes target and probe hybridization as well as target and probe dissociation when washing the arrays. First results have already demonstrated that especially the target and probe dissociation during the washing steps with stringent wash buffers have the strongest influence on the amount of signal that is detected later. Another relevant finding has been that non-specific cross-hybridization has a much stronger effect on the accurate quantification of absolute gene expression intensities than any three-dimensional folding of the target that is being hybridized to the array[38].

3.1.1.3 Expected changes in mRNA levels

To avoid getting sidetracked by spending time on data from a microarray experiment that did not result in the expected biological response, it is helpful to have various controls. One of these controls are mRNAs that are expected to be altered by the experiment. For example, when a given stimulus increases the stress level of an animal, an indicator of such increased stress can be checked by looking at the gene expression measures of stress inducible genes

---

[1]The most commonly used approach involves physical models in the form of Langmuir derivatives. Such models are used to describe first-order chemical reactions (here the binding between target molecules and the probes on the arrays). Typically, the modelling assumes an equilibrium at the end of the hybridization process. At this point a balance has been reached between the number of targets binding to probes and the number of targets dissociating from the probes. However, there are some issues with this approach, e.g., the Langmuir model requires probes to saturate at the same level, which is clearly not the case with Affymetrix arrays: usually low-affinity probes saturate first[37].

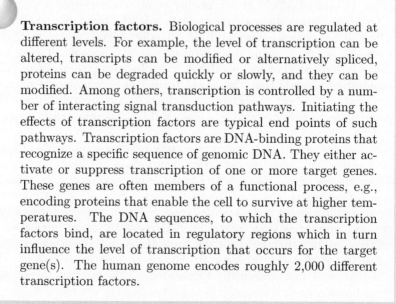

Transcription factors. Biological processes are regulated at different levels. For example, the level of transcription can be altered, transcripts can be modified or alternatively spliced, proteins can be degraded quickly or slowly, and they can be modified. Among others, transcription is controlled by a number of interacting signal transduction pathways. Initiating the effects of transcription factors are typical end points of such pathways. Transcription factors are DNA-binding proteins that recognize a specific sequence of genomic DNA. They either activate or suppress transcription of one or more target genes. These genes are often members of a functional process, e.g., encoding proteins that enable the cell to survive at higher temperatures. The DNA sequences, to which the transcription factors bind, are located in regulatory regions which in turn influence the level of transcription that occurs for the target gene(s). The human genome encodes roughly 2,000 different transcription factors.

BioBox 3.2: Transcription factors

such as the immediate early gene transcription factors (IEG-TFs)[39],[40] after the microarray data are obtained (see also BioBox 3.2).

3.1.2   Sample

When running gene expression microarrays on the Affymetrix platform, there are predominantly two types of experiments: one type involves the use of cells that were grown in cell culture disks in the laboratory (*in vitro*) and the other type studies gene expression in cells obtained from a living organism, e.g., human biopsies, animal tissue samples, blood, etc. (*in vivo*).

3.1.2.1   Selection of appropriate sample/tissue

Without sufficient background knowledge as to where gene expression alterations are most likely to occur, it can be very difficult and costly to make the right choice when having to empirically determine the correct tissue that should be obtained. This is especially true when doing experiments in higher organisms such as mammals with a multitude of different tissues. When conducting experiments that involve treatments leading to, e.g., alterations in the brain, it is often not advisable to look at total brain mRNA extracts. Small and at times also large alterations will get diluted due to the specificity of where a gene is expressed in the brain. The expression of the gene Transthyretin in the brain of mice illustrates this problem. Transthyretin is expressed in only very distinct areas (see Section 3.1.2.6). If one were to use mRNA from total brain extracts the abundance of this specific mRNA would get heavily diluted in comparison to the other transcripts.

Therefore, running a pilot experiment can avoid surprises if one is not certain about the right tissue. Scientists often consider varying the intensity of a treatment or the treatment duration. However, the choice of the right tissue is often ignored. This is mainly due to scientific background and knowledge that one has at the time of the experiment about a disease or the location where a treatment has an effect. It can happen though, that the intended effect is much more clearly visible in a different tissue. This can be seen, e.g., in experiments that study a certain brain region, while a different region gets in fact more affected by the experiment.

Furthermore, educated guesses of where else to look can be quite interesting, e.g., when running experiments with drugs that induce antidepressant effects in the brain, it can be informative to look at blood expression levels but possibly also for effects on the gastrointestinal system as many people taking antidepressants report gastric distress.

3.1.2.2 Sample types

Tissues The expected yield of total RNA varies largely between different organs. Furthermore, nucleases (enzymes that digest nucleic acids such as DNA and RNA) are present in sample tissues to various degrees which in turn can have a negative effect on the RNA quality. Therefore it is advisable to assess the quality and quantity of a given tissue in a small pilot experiment. An appropriate starting point is a $2 \times 2 \times 2$ mm cube of tissue. It should weigh approximately 5 mg and should yield around 10 μg of total RNA as a general rule of thumb. Table 3.1 has a more detailed overview of what yields to expect per tissue type.

Using a small tissue cube also ensures that the block is immediately frozen when submerged in liquid nitrogen. This way artifacts due to nuclease activities or potential cold shock reactions of the cells can be avoided. While it is perfectly feasible to extract RNA from tissue samples immediately, oftentimes the sample will be processed later. It should be snap frozen in liquid nitrogen within 30 minutes after dissection. Otherwise proprietary RNA stabilizing solutions such as RNAlater®[2] can provide an alternative.

Tissue samples are typically stored at -80°C to prevent RNA degradation. An alternative is the storage of samples that have been homogenized together with a denaturant that inactivates RNases, at -20°C.

---

[2]Registered trademark of Applied Biosystems, Inc., Foster City, California.

Blood consists primarily of two components: An aqueous solution called blood plasma and blood cells. While the plasma contains various molecules such as proteins (e.g., antibodies detecting foreign molecules), nutrients (e.g., sugars), hormones, waste products, etc., it also contains different types of cells. The vast majority are red blood cells (also referred to as Erythrocytes) which carry oxygen inside them captured in a molecule called Hemoglobin. Of relevance to microarray research is the fact that red blood cells do not contain a nucleus. Therefore they have no DNA and show no transcriptional response to external stimuli. However, they do contain large amounts of mRNA encoding the protein portion of Hemoglobin.[a] The smaller portion of blood cells comprise the white blood cells (termed Leukocytes) that are part of the immune system and the platelets (Thrombocytes) that ensure that bleeding stops after injury by creating a clot composed of the protein fibrin and red blood cells.

---

[a]Hemoglobin is a protein that contains iron. The red color is typically seen outside the body where the iron is rapidly oxidized ("rusted") when exposed to air. As oxidized iron is essentially rust, blood has the same typical smell and color of rust.

BioBox 3.3: Blood

Peripheral blood When studying gene expression of human tissues, it is at times difficult to obtain primary tissues. However, peripheral blood is a tissue that is often easy to obtain and thus gets increased attention in the search for molecular diagnostics. The quality of the RNA and its corresponding data is largely influenced by the sample collection, sample storage conditions and the subsequent processing procedures. Especially RNA degradation has been shown to affect the quality of data obtained from blood as both ribosomal and mRNA are quickly degraded in whole blood when not preserved in some form[41].

While RNA can be isolated from whole blood or from peripheral blood mononuclear cells (PBMCs), often PBMC are isolated as they contain the most transcriptionally active cells in the blood. Due to the large differences in data obtained from gene expression profiles of RNA from PBMCs vs. RNA from whole blood, one should not use them interchangeably in the same study.

There are a number of different protocols available for the isolation of blood cells for RNA analysis. When deciding on a certain approach, one needs to keep in mind that all procedures will affect the detected gene expression to some extent and that the various procedures will also differ in how they affect the measured mRNA levels.

Another point of attention are possible confounding effects when detecting gene expression differences that are actually caused by differences in cell counts for different cell types between individuals. A possible solution is to check for cell counts using, e.g., flow cytometry in cases where such confounding cannot be excluded or when sample sizes in the experiment are very low.

Whole blood One commonly applied system for collecting whole blood is the use of PAXgene™[3] tubes. These tubes contain a RNA stabilizing agent which protects RNA molecules from active degradation by enzymes that are referred to as RNases. At the same time such stabilizing solutions minimize any further changes in gene expression. Each PAXgene™ tube is used to collect 2.5 ml of human whole blood. The first step in the procedure sediments nucleic acids via centrifugation. Afterwards the pellet is washed and proteins are digested using the enzyme proteinase K. An additional centrifugation through a specific spin column is done to homogenize the cell lysate. Subsequently the supernatant containing the nucleic acids is applied to a RNA spin column where only RNA can bind. The nucleic acid is washed in the column and any bound DNA is removed by an enzymatic digestion with DNase I. After the wash steps, RNA is eluted. A major advantage of this procedure is the simplicity of the protocol. Blood can be collected by persons who do not need to have experience in handling samples for RNA analysis.

It is important to highlight that RNA yields from whole blood are highly donor-dependent and therefore individual yields can vary between 2 and 20 μg of total RNA per collection tube.

When using total blood for gene expression analysis, one has to consider that reticulocytes (immature red blood cells (RBCs)), while representing only 0.5-2% of the RBCs, can contribute approximately 70% of the mass of mRNA in total RNA, of which globin mRNA is the major RNA. Depending on the protocol used for the synthesis of the labeled target, a step to reduce globin mRNA is included. By doing so, one can detect approximately 10%-15% more genes (when looking at the percentage of probesets called "present" by the MAS 5.0 detection call algorithm), which are primarily low abundance transcripts[42]. In other words, the overabundance of globin mRNA bears the risk that transcripts that are of low abundance are not detected.

Affymetrix provides a globin-reduction protocol to drastically lower the abundance of globin mRNA in the RNA sample. Companies like NuGEN™[4]

---

[3]Registered trademark of PreAnalytiX GmbH, Hembrechtikon, Switzerland.

[4]Trademark of NuGEN Technologies, Inc., San Carlos, California.

have brought kits to the market such as the Ovation Whole Blood Solution[TM][5] which does not need a reduction of globin mRNA, but potentially gives rise to other challenges such as contamination with genomic DNA and a shift in specificity and sensitivity due to the use of DNA instead of RNA as hybridization material.

Isolation of peripheral blood mononuclear cells (PBMC) RNA isolation from PBMCs has a major advantage as the RNA is free from globin mRNA. This in turn reduces the background caused by globin RNA when profiling mRNA from whole blood. The higher sensitivity of the microarrays generally leads to larger numbers of detectable transcripts as mRNAs that are expressed in the cell in low numbers can also be detected.

When isolating PBMCs one harvests white blood cells, which are the most transcriptionally active cells of all blood cells. For this technique, blood is layered on top of Ficoll-Paque[TM][6] and forms a density gradient with different layers during a centrifugation step. One layer will then contain the white blood cells (lymphocytes and monocytes) while other layers will contain the red blood cells (erythrocytes) and one other type of white blood cells (granulocytes). A drawback of this approach is the long processing time of typically more than one hour and possible temperature changes, which can introduce changes in gene expression as the cells are still intact and able to transcriptionally respond to alterations in their environment. One can typically see changes in gene expression linked to biological processes such as metabolism, hypoxia, cell cycle regulation and apoptosis[43]. Therefore PBMCs should be isolated within hours of blood collection.

Furthermore, this procedure requires trained personnel to avoid differences in sample quality due to lack of experience. Irrespective of the skill of the experimentor, it needs to be assessed how consistently and reliably RNA is obtained between samples of the same experiment.

Human postmortem brain Gene expression profiles related to psychiatric disorders can only be studied using postmortem brain samples. Therefore it is essential that the data obtained from such samples is as closely related to the situation in living brain cells as possible. Besides assessing confounding factors such as medication (e.g., neuroleptics), smoking, gender, age and diet, the sample quality itself needs to be controlled: RNA quality, agonal factors (e.g., coma, hypoxia, hyperpyrexia at the time of death), length of postmortem interval (PMI) or potential thawing of samples in the absence of RNA stabilizing agents could have profound effects. While there are reports that the agonal state has much more influence on the RNA integrity than gender or age[44], this assessment has also been recently contradicted. It seems that the quality of the extracted RNA depends primarily on the technology

---

[5]Trademark of NuGEN Technologies, Inc., San Carlos, California.
[6]Registered trademark of GE Healthcare Bio-Sciences, Uppsala, Sweden.

while the agonal state and the pH of the brain samples does not seem to have influence[45]. Furthermore, it has to be made clear that it is currently not certain whether the observed variability between samples is more likely to be caused by the aspects mentioned above or by tissue heterogeneity (see also Section 3.1.2.6 on dissection artifacts in mouse brain samples).

FFPE Tissue processing done routinely at, e.g., hospitals has produced many large banks of samples that are formalin-fixed and embedded in paraffin (FFPE). While such tissue banks are very interesting for the researcher to conduct retrospective studies, e.g., what difference could be discovered between patients who responded to a chemotherapeutic and those who did not respond, they often pose a set of problems:

1. The age of the sample and accordingly the level of degradation of RNA in the samples.

2. Differences in fixation protocols. The process of generating FFPE material is more than 100 years old, but there has never been a standardized procedure to preserve tissue histology or molecules such as DNA or RNA. Furthermore, the use of Formalin as a fixative leads to crosslinking of DNA as well as RNA, which in turn prevents the isolation of nucleic acids that are longer than 100 or 200 base pairs and are often chemically modified. RNA is further exposed to degradation due to cellular processes and tissue autolysis.

3. Differences in subsequent RNA extraction and amplification protocols. RNA amplification efficiency is largely influenced by the quality of the RNA (e.g., level of degradation and level of chemical modification).

All these factors have an impact on the RNA quality, introduce extra variation into the data and make generalizable conclusions more difficult. This is especially true for RNA, which is in contrast to DNA even less stable in FFPE material[46]. Therefore, according to a study by Penland et al.[47], only a minority of FFPE samples will yield RNA of a quality that is sufficient enough for microarray analysis.

Cell lines While offering the scientist the best quality RNA and arguably the easiest way to control the experimental conditions, the use of cell lines as experimental units has clear limits when trying to generalize the findings to whole organisms (see BioBox 3.4). Of course, the closer the cells of a cell line resemble the cells in a human tissue, the more likely it is that experimental results will translate well to the human situation. However, while primary cells resemble the human cells better than cells from continous cell cultures, they typically lose their tissue-specific differentiated phenotype quickly once brought into culture. A fully differentiated phenotype is also often linked to limited proliferation so that it is difficult to expand a culture of primary cells in the lab. Taking further into account a high variability between batches of primary cells, the high cost price linked with the difficulty to obtain primary cells from humans, it becomes clear that continous cell cultures, while even more limited than primary cells, tend to be the model system of choice for many basic research questions. The fact that continuous cell cultures often acquire substantial alterations in their chromosomal composition in the course of immortalization needs to be kept in mind when attempting to generalize study results. Such alterations can lead to various functional characteristics that are no longer related to the original primary cells.

Running gene expression profiling experiments on samples from cell line cultures, the use of 6 well cell culture plates typically results in sufficient amounts of total RNA per well. The yields are typically sufficient for standard Affymetrix protocols for both 3' focused transcript arrays or whole transcript arrays. When a microarray experiment is done on cell culture samples, it is advisable to reserve a separate incubator for the whole experiment to avoid fluctuations in the temperature. When harvesting the cells, immediately aspirate the medium completely and disrupt the cells by adding a lysis buffer. There is usually no need for mechanical or enzymatic disruption of the cells. Swirling the cells for 10 min on a plate shaker at moderate speed is mostly sufficient.

Even though it should be clear that aspects such as using cell cultures at a high number of passages, cellular cross-contamination, or the percentage of confluency have an impact on gene expression, it regularly occurs that scientists overlook one of these issues and later face experimental results that are nonconclusive or even misleading. For example, every reseeding of cells into new medium[7] puts a selection pressure on the cells that can result in genetic drift, e.g., selecting for fast growing cells. In other words, cells will adjust more and more to their new artificial environment, thereby loosing their original characteristics such as their constitutively active pathways. Current data estimate that more than 18% of cell lines of cell repositories are either misidentified or contaminated[48]. Therefore it is advisable to either obtain a cell line from one of the major cell line banks or assess the quality of a cell line by authentication tests and checks for possible contaminations. Furthermore, the passage number should be kept as low as possible and the level of confluency to which cells are grown in the course of an experiment should be tightly controlled.

3.1.2.3 Sample heterogeneity

Even though many organs are well defined, they often do not contain cells of one single type. For example, the brain which contains neurons (the nerve cells) that are often thought to be the "typical" brain cell, is made up of up to 90% of glial cells that do not carry any nerve impulses.

---

[7]Medium refers to the liquid containing all necessary nutrients in which cells grow artificially.

Cell culture. There are primarily two types of cell cultures: *primary cell cultures* which are derived from tissues and *continous cell cultures* which are in essence transformed cells such as cancer cells. A main characteristic of primary cell lines is that they have a limited life span when cultivated in cell culture, while continous cell lines theoretically divide indefinitely. Cultivating cells is done in plastic flasks or plates with multiple wells[a] that contain a medium with all required nutrients for the cells to live, grow and divide. For cell culture-based experiments typically a certain number of cells are brought into the culture medium (referred to as "seeding density") and are incubated at 37°C. Once the cells have grown to cover roughly 80% of the surface of the flask (referred to as "80% confluency") the cells are harvested. For microarray experiments one would at this moment remove the medium and lyse the cells using a lysis buffer that simultaneously protects the RNA from degradation and inactivates all RNA degrading enzymes. If cells are grown in culture to produce larger amounts of cells, the cells will be grown to roughly 80% confluency and then seperated into two or multiple new flasks or wells. This process is called "splitting." Every time a split is done, the number of passages is increased by one. The passage number therefore indicates how often a given cell line has been split into new sub-cultures. To avoid genetic drift, passage numbers should not get too high.

---

[a]Well is the technical term used for specifying the small "tubes" that are arranged on a grid of a flat plate. Each well can typically hold several milliliters of liquid. They can be either circular or square. Cells are grown adherently in these multiwell plates.

BioBox 3.4: Cell culture

A similar situation can be seen when looking at tumor tissues: the initiation and progression of human tumors typically result from a series of mutation and selection events. Mutations occur frequently in genes of pathways involved in regulating tumor progression (e.g., oncogenes[8]). Once a cell has become tumorgenic and is not identified by the immune system, it can divide in an uncontrolled way. As it also tends to accumulate even more genetic alterations and even gross chromosomal abnormalities, the cells that are most fit for the environment are being selected and grow fastest. Furthermore, in the center of the tumor the supply with nutrients typically diminishes leading to cells that have stopped dividing or are already dying. The differences between cells on the genetic level and between environmental conditions result in diversity in between cells and to an even larger diversity between tumors of different patients of the same tumor type. In other words, cells with differences in their genetic alterations cause regions of the tumor with cells of one set of mutations while in a different region the cells might have acquired a different set of mutations. Furthermore, tumors are known to consist of tumor tissue and non-tumor tissue.

Therefore one has to decide whether this sample heterogeneity can be ignored and RNA is extracted of the whole tumor or whether one needs to make sections of the tumor and assess the heterogeneity of the tumor with *in situ* histochemistry of slices out of these sections. This is of course most relevant for tumors that are larger in size.

As an example to demonstrate when it can be acceptable to ignore the heterogeneity is the following: studying the dependence of a tumor on a single oncogenic pathway (a phenomenon called "oncogene addiction") could be done on heterogeneic material as it has been shown that this dependence occurs despite a large diversity in genetic alterations[49].

For experiments where one decides to reduce the complexity level, dissecting out cells of one type in a heterogeneous tissue is often the most obvious first choice. This enables the researcher to analyse more homogeneous biological samples. However, it relies on experience to reproducibly remove the same part of a tissue across samples and bears the risk of dissection artifacts (see Section 3.1.2.6).

A more sophisticated approach involves a technology called "laser capture microdissection (LCM)." The procedure makes use of a laser to cut out certain cells from a section of a tissue block. While LCM offers the potential to study even single cells, the procedure involves several manipulation steps (including the preparation of sections, staining of cells, etc.) which in turn can have a negative impact on RNA quality. Furthermore, the amount of RNA obtained from LCM samples is of course much smaller than from whole tissue blocks. This requires often extra amplification steps that can introduce biases in the data (see Section 3.1.2.9).

---

[8]Oncogenes are genes that cause a cell to survive indefinitely and proliferate abberantly.

Xist. Female mammals have two X chromosomes while male mammals have only one X chromosome and some extra gender-specific information on the Y chromosome. Since all genes on the X chromosome are twice as often present in females, there needs to be a mechanism that avoids the fact that also twice as much of the encoded protein would be made in females. At the basis of this mechanism lies the gene Xist. This gene encodes for a large transcript that is not a mRNA (i.e., non-coding = not translated into protein). Rather the RNA itself gets stably associated with one of the two female X chromosomes and completely inactivates it. Experiments have shown that this inactivation is strictly limited to the chromosome where the Xist transcript is made. In other words, when experimentally copying the Xist gene to a different chromosome, this chromosome will get inactivated in a similar fashion.

BioBox 3.5: Xist

3.1.2.4 Gender

Even though many scientists do not expect to see an effect of gender on their experimental outcome, there are of course a number of genes that are clearly gender specific. Dataset HD is using a population of both female and male mice. Looking at a data-driven grouping of samples using spectral map analysis (see Figure 3.2) one can see already the impact of gender on the overall variability in the data as the gene Xist is highlighted. Figure 3.1 shows the intensities for the probes of the probeset. One can clearly see that samples GSM197115, GSM197116, GSM197120, GSM197124, and GSM197125 are from male mice as they do not express the gene.

3.1.2.5 Time point

Upon exposing an organism or cell lines to a stimulus, changes in gene expression occur as the organism adapts to the altered environment. These adaptations are dynamic, e.g., adding sugar to the medium of a yeast culture will turn on the enzymes that can utilize the sugar until it has been depleted. Then the genes for making the involved enzymes are turned off again.

This example illustrates that choosing the right time point for the termination of the experiment (i.e., the snapshot of gene activity at this moment) has to be chosen carefully. Especially when one is not familiar with the biological setting, running a time course pilot experiment is essential.

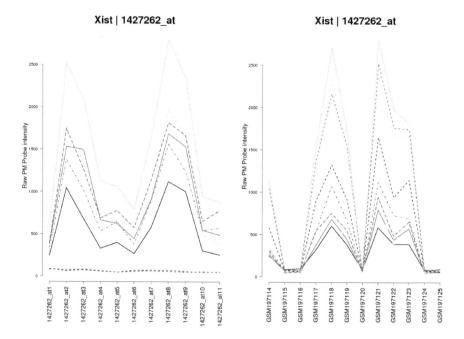

FIGURE 3.1: Gender specific gene Xist. This probeset detects RNA levels for the gene "inactive X specific transcripts" (Xist). This gene is a marker for gender as male mice have no Xist RNA while the gene is actively transcribed in female mice. From the right graph samples GSM197115, GSM197116, GSM197120, GSM197124, and GSM197125 can be identified as male mice.

Furthermore one also needs to realize that it is not always possible to infer the choice of the correct time point from results based on protein experiments. Changes on the protein level are typically taking place after changes in gene expression have occurred.

Another factor that can be of relevance in this context are biological experiments in which one would want to study a certain effect in multiple species or cell lines. For example, cell lines can have very different growth characteristics and also very different genetic backgrounds. When one looks at the effects of a treatment in multiple responsive cell lines, the desired gene activity alterations can take place at very different times. In other words, doing a pilot experiment in one cell line to determine an optimal time point for a gene expression readout is likely to be inaccurate in other cell lines.

3.1.2.6 Dissection artifacts

A multicellular organism relies on cells that have specialized to form certain tissue types and organs. When studying gene expression, one can differentiate between genes that are expressed at constant levels in ideally all cells (these are often referred to as "housekeeping genes") and genes that are predominantly expressed in one or only few tissue types. In contrast to the housekeeping genes are those genes typically encoding a gene product with a very specialized function. The expression of these genes tends to be limited to one cell type.

It is this tissue specificity of some genes that can lead to problems when the experimentor tries to dissect out only cells from a single tissue type. Even though the experimentor will of course be as accurate as possible during the dissection, the sample still might contain trace amounts of material from neighboring tissues. In turn these contaminations can influence the measured gene expression intensity, especially if there are genes that are exclusively expressed in the neighboring tissue. Figure 3.2 shows an example of such an artifact detected using spectral map analysis. This example was generated using the dataset HD on the striatum of mouse brains (see also Figure 3.3 from the Allen Brain Atlas[50],[51]). Another example of a dissection artifact is the dissection of colon samples from mice. Here one can sometimes detect in the gene expression data that surrounding tissue from the pancreas have been dissected out as well. One example dataset that shows this potential problem is GSE10740. As mouse colon samples will occasionally have some remaining pancreatic tissue, this artifact can be detected by a completely different pattern for some pancreas specific genes such as Amy2, Ela3, Prss2, Pnlip, etc. Figure 3.4 shows the gene expression data for the three genes mentioned above. One can clearly see that sample GSM271147 contained a substantial amount of surrounding pancreatic tissue.

3.1.2.7 Artifacts due to animal handling

When using laboratory animals for conducting experiments that are subsequently analyzed by microarrays, one needs to establish uniform procedures for the handling and the treatment of the animals to ensure the reproducibility of the generated data. Randomly assigning animals to the different treatment groups avoids detecting effects which cannot be clearly derived from the treatment and might have been caused by, e.g., assigning the oldest animals to the treatment group and the youngest animals to the control group.

Especially when experiments are split up over time to compare multiple treatments or larger numbers of animals, there are a number of factors that need to be controlled. Among others the number of animals per cage vs. single-caged housing, diet, gender, age, length of fasting, time of day (also referred to as "diurnal cycle"), and anesthesia used when drawing blood have been shown to profoundly affect biological measurements[52].

For example, even though many experiments involving laboratory animals will often be unrelated to the diet of the animals and during the experiment the composition of the food will typically not be altered, it can still be of relevance to know that some dietary elements can have an influence on the gene expression of organs like the liver or the lung. Especially the estrogenic

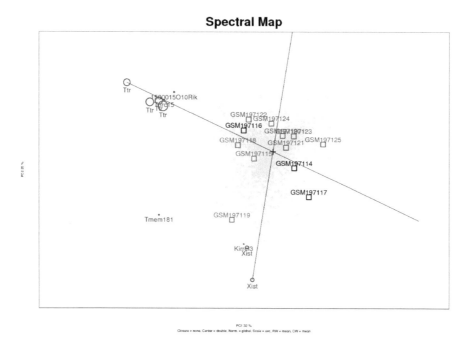

FIGURE 3.2: Example of a dissection artifact. The striatum of the brain of mice treated with different levels of a compound were dissected out and profiled on microarrays. Along the line connecting the transthyretin (Ttr) gene with the origin (black cross in the center of the graph) samples vary with respect to the extent of the dissection artifact. This seperation is caused by the inclusion of different amounts of neighboring tissue and thereby by different levels of the marker gene transthyretin as detected by four different probesets for transthyretin on the whole genome mouse array 430v2 (four circles on the left). Furthermore one can detect another influence on the data: two probesets detecting the gene "inactive X specific transcripts" (Xist) also separate the samples. This gene is a marker for gender as male mice have no Xist RNA while the gene is actively transcribed in female mice. This identifies a mixed population of male and female mice in the experiment.

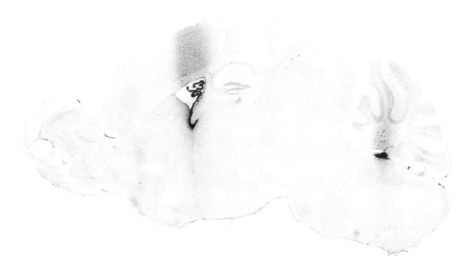

FIGURE 3.3: The *in situ* hybridisation of a probe against transthyretin shows nicely the very discrete expression of the gene. In the areas where this gene is expressed, the level of expression is so high that it has led to some artificial bleeding of the stain into the surrounding areas. Depending on the dissection, more or less of the neighbouring tissue surrounding the striatum will be included thereby including at times large quantities of transthyretin mRNA. Image (Ttr_93 (position 2325)) from the Allen Brain Atlas.

activity of the diet and contaminations with toxicants have been identified to have an influence on the measured gene expression levels[53]. One example is the varying concentrations of methylmercury that are likely introduced into non-purified diets via the addition of changing concentrations of mercury in fishmeal[54]. It is therefore advisable to use defined diets (also referred to as "purified diets") as the effect on gene expression will be more limited and more consistent[53].

Another potential source of unintended variability can be the killing and dissection of the animal. Affected are especially experiments that are likely to involve alterations in stress genes or whenever the dissection and isolation of a particular tissue is time consuming. Of course the extent of such alterations varies between animals.

3.1.2.8 RNA quality

When running gene expression profiling experiments, one of the most critical steps in the lab is the RNA extraction. Obtaining high quality, intact RNA in sufficient amounts is essential. Another important aspect that should be

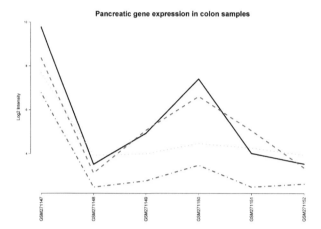

FIGURE 3.4: Dissection artifact observable in mouse colon samples. Each line represents the gene expression intensities for four different genes in six colon samples from GEO dataset GSE10740. Especially the first sample GSM271147 contained some surrounding pancreatic tissue after the colon sample was dissected out. This can be seen by the high expression level of the four genes Amy2, Ela3, Prss2, and Pnlip which are either non-detectable or low expressed in colon but are strongly expressed in pancreatic tissue.

monitored is the shipment of samples. The conditions that are used (are the samples pelleted cells on dry ice or are they already lysed or are they tissue samples submerged in a RNA stabilizing agent, etc.?) have an influence on the integrity of the RNA. Linked to the shipment conditions are also the storage conditions of either the undisrupted samples or the isolated RNA.

When looking at the RNA prior to running a microarray experiment, there are different characteristics that can be assessed. While it might be possible that, e.g., a certain procedure results in a 28S:18S rRNA[9] ratio of only 1.8 (see below), the variability between samples of the same experiment and across experiments needs to be kept to a minimum. Large deviations often indicate insufficiently optimized lab procedures. There are exceptions, e.g., RNA from total human blood varies largely between donors, but for many tissues the

---

[9]18S rRNA and 28S rRNA refer to ribosomal RNAs that are part of the ribosomes in a cell. Ribosomes are the complexes where proteins are being built based on the mRNA template. Since ribosomes are so important, the ribosomal RNAs represent by far the bulk of the total RNA in a cell. Due to their different sizes one can use the difference in abundance between the 18S rRNA and the 28S rRNA as a measure of RNA quality.

The S in 18S or 28S represents Svedberg units which characterize the sedimentation rate of a particle type in ultracentrifugation. The shape and the mass affect the sedimentation rate whereby bigger particles generally have larger S values.

Gel electrophoresis is a molecular biology technique that separates molecules of a certain class (e.g., RNA, DNA or proteins) according to size. These molecules are initially present in a complex mixture that is obtained by disrupting cells, e.g., from a biopsy. The separation is done by applying an electric current to a gel matrix made of, e.g., agarose or poly-

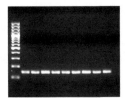

Agarose gel

acrylamide. The current makes the charged molecules migrate through the gel. As small molecules will be less hindered by the gel matrix, they will travel further. Subsequently the molecules are stained and visualized generating an image as shown in Figure **??**. To be able to assess the size of an unknown molecule, the first lane contains a mix of molecules of know sizes. For the analysis of RNA oftentimes a 2% agarose gel is used under denaturing conditions so that the RNA molecules do not form any secondary structures that could have an effect on the linearity between distance travelled in the gel and size of a RNA molecule.

BioBox 3.6: Gel electrophoresis

level of consistency between RNA extractions is the best indicator for good microarray starting material.

Ratio 28S:18S rRNA The ratio between the amount of 28S rRNA and 18S rRNA is typically used as an indicator for RNA quality. While this ratio should be ideally 2, there can be some deviations depending on the tissue type. There should certainly not be any bands visible between the two ribosomal bands and below when looking at a sample of a RNA preparation via agarose gel electrophoresis (see BioBox 3.6).

RNA storage Depending on the sample, alterations in the gene expression profile can be detected when a tissue sample is kept at room temperature or 37°C, even though this might have no impact on RNA integrity. Especially genes concerning responses to stress and genes known to be immune cell markers tend to get differentially expressed if the tissue sample is not snap frozen after dissection. However, RNA from cells that have been snap frozen will rapidly degrade when thawed in the absence of RNA protecting agents. In other words, the longer a sample is unprotected after thawing, the stronger an effect on RNA quality can be seen. This time-dependent effect is much more drastic than the time prior to freezing the samples. The destruction of the cellular structure which in turn makes RNA more accessible for degradation is the cause for this phenomenon. Autolysis therefore seems to take much longer. As a consequence of RNA degradation, the average length of the obtained products generated during the labeling of the RNA decreases[55].

An alternative to snap freezing are RNA stabilizing agents such as RNAlater. Such aqueous agents avoid the necessity to immediately process the samples as they offer RNA stability of up to one week even at room temperature. As it can be very difficult to further process tissue blocks prior to RNA extraction that have been submerged in a RNA stabilizing agent, it is important to ensure that tissue blocks are no larger than 5 mm in any single dimension. Products like RNAlater will however not improve the quality or yield of subsequent RNA extractions.

In general, RNA stored for short term should only be dissolved in RNase-free water. For somewhat longer storage one should consider RNas-free TE buffer (buffer made from Tris and EDTA) pH 7.0 as it reduces hydrolysis of RNA through chelation of divalent cations by EDTA and through resistance to pH changes by the Tris buffer. These samples should be stored at -80°C in aliquots. Repeated freeze-thaws should be kept to a minimum. RNA is at least stable for one year at -80°C under these conditions. For long term storage, RNA samples could be stored at -20°C in ethanol.

Degradation   RNA can be easily degraded by the enzyme RNase which is ubiquitously present in living organisms. Degradation is therefore a major concern when working with mRNA. The enzymatic activity of RNase which is present in all cells and also on the surface of human fingers results in increasingly smaller fragments that can be seen on the image taken from a gel electrophoresis run or from a bioanalyzer run (see BioBox 3.7 and Figure ??). The image will often also lack discrete bands for the 28S rRNA and the 18S rRNA (see Figure 3.5).

Optimal methods for sample handling are necessary to keep the RNA intact. This includes the use of RNase-free reagents, consumables, equipment and glassware. Furthermore, it is adviseable to dedicate work spaces and equipment specifically for RNA isolation and other RNA-related work.

RIN number   The RNA Integrity Number (RIN) is the output generated by an algorithm to provide the scientist with an objective assessment of the integrity of a given RNA sample. The algorithm uses data from an analysis run of a RNA sample on the Agilent Bioanalyzer. Resulting values range from 1 indicating completely degraded material to 10 for high quality RNA with no

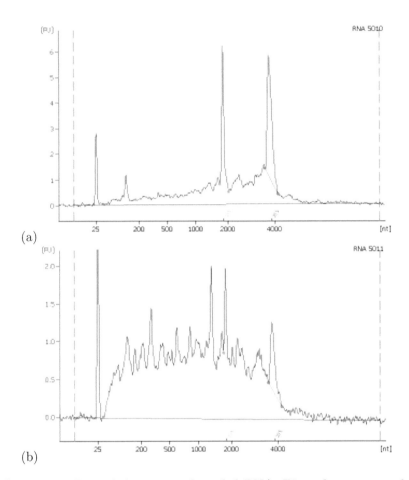

(a)

(b)

FIGURE 3.5: Degraded vs. non-degraded RNA. Bioanalyzer traces showing an example electrophoretic trace for non-degraded, high quality RNA (a)
and degraded RNA (b). The two dominant peaks that can be seen for non-
degraded RNA indicate the amount of 18S and 28S rRNA that is present in
the sample. As RNA degrades, these peaks together with all other RNAs
(including the mRNAs) get reduced to a mix of smaller, fragmented RNA
pieces.

Bioanalyzer. The 2100 Bioanalyzer from Agilent Technologies is a microfluidics-based instrument for determining the quantity and quality of a RNA sample. Major advantages of this system over traditional gel electrophoresis are the smaller amounts of RNA needed per analysis and a reduction of manual steps needed to perform the assessment. Part of the analysis of the data obtained by the Bioanalyzer is the calculation of the RIN number (RNA integrity number). This single number has become a standard tool to judge the overall quality of the RNA. The system uses an intercalating fluorescent dye that monitors fluorescence between 670 nm and 700 nm.

Bioanalyser

BioBox 3.7: Bioanalyzer for RNA quality assessment

detectable degradation products. Instead of the classical visual inspection of a gel electrophoresis image, the algorithm incorporates information from both the ribosomal RNA bands (18S/28S) and the data of the whole electrophoretic trace. The advantage of this approach is a much more objective indication of RNA quality which enables the scientist to standardize the assessment RNA quality across experiments. In general all samples with a RIN number of 8 or higher are of sufficient quality. Samples should not have a RIN number below 6. For samples with a RIN number between 6 and 8, it is advisable to look at the variability of RIN numbers for all the samples in an experiment. In case all samples have RIN numbers between 6 and 7, then this could indicate a limit of the lab procedures. One can either try to optimize the protocols further or continue with the samples taking into account possible loss in signal. However, it is most important that the sample quality is consistent. Single samples with a low RIN number should rather be removed from further analysis.

3.1.2.9 RNA quantity

One commonly applied kit for RNA isolation is the RNeasy kit from Qiagen. Table 3.1 gives an overview of typical yields one can expect for a few tissue types. In some circumstances, e.g., when isolating cells via LCM (see Section

TABLE 3.1: RNA yields for tissue samples and cell culture samples.

Tissue	Yield (μg)	Starting Material (mg)
Bone	0.01	1
Brain	1-2	1
Heart	4	1
Kidney	2-4	1
Liver	5-10	1
Lung	2	1
Placenta	1-4	1
Skeletal muscle	1-2	1
Spleen	5-10	1
Cultured epithelial cells	10-15	1
Cultured fibroblast cells	5-10	1

3.1.2.3), when looking at sorted cells or when looking at RNA from fine needle aspirates, it can be necessary to amplify the amount of input material for microarray analysis.

Small amounts of total RNA can be amplified via an extra round of amplification. The standard protocol uses total RNA and converts it into cDNA enzymatically using reverse transcriptase. Afterwards there is a linear amplification step whereby the cDNA is used to make RNA again[10]. This step amplifies the material so that one original copy of a transcript results in multiple copies of the same sequence[56]. The generation of this material is done in the presence of altered ribonucleotides[11] that have a special modification onto which, after the hybridization and washing of the arrays, a flurophore is attached.

Instead of applying this standard protocol, it is possible to use the RNA that is made after the amplification as starting material for an extra round of cDNA synthesis which is then again used as a template for the generation of RNA[12]. This procedure allows one to obtain gene expression data from less

---

[10]This process makes use of primers that have an oligo(dT) sequence as well as the promotor sequence for the T7 RNA polymerase. Once the cDNA is synthesized, the promotor sequence is the starting point for the RNA synthesis. As the polyA tail of a mRNA is at the 3'-end of a transcript and the synthesis efficiency of the reverse transcriptase is not 100%, a bias towards the 3'-end is introduced.

[11]The ribonucleotide mix used for this step contains ribonucleotides with a biotin group. Using a conjugate of streptavidin with phycoerythrin the staining process makes use of the strongest non-covalent biological bindings that are known: the binding of biotin to streptavidin has a dissociation constant $K(d)$ of 4×10^{-14}M. This way all biotin labeled nucleotides receive a flurophore, which is excited during scanning. Accordingly the relative abundances originally present in the sample are preserved.

[12]During the first RNA amplification step unlabeled ribonucleotides are used. The second cDNA synthesis is primed with oligonucleotides that contain random sequences, as the synthetic RNA that is used in the first amplification step does not contain a poly A tail anymore.

than 10 ng of total RNA. However the use of random hexamers in the second cDNA synthesis step clearly increases the 3' bias in the data: probes that are located close to the 3'-end of a transcript will give much stronger signals than probes located closer to the 5'-end. Therefore, it is worthwhile to check via degradation plots the impact on the signal (see Figure 3.6). Besides the signal intensity bias, one can also generally expect to see a decrease in the number of detectable genes ("present" calls) as well as an increase in noise.

Even though there seems to be less of an impact on the results it can also be interesting to investigate whether specific summarization techniques could help to enhance the results such as sRMA[57] which was specifically developed in the context of amplification and the impact on the data (see Section 4.1.5).

Over the last years the required amounts of starting material have been reduced substantially. While it was initially necessary to have up to 15 μg of total RNA, it is now sufficient to start with 1-2 μg of total RNA. With the recent introduction of the IVT express kit, this has now been lowered to 50-100 ng of total RNA. Only experiments that result in smaller yields of total RNA still require an extra amplification step.

 Do not mix protocols. When running a microarray experiment that is likely to result in further microarray experiments, it is important to decide on one protocol for the whole study. The biases that are introduced via different protocols make it difficult to compare results across experiments that make use of different protocols. This advice is not limited to standard protocols vs. protocols with an extra round of amplification. It is also relevant if one were to consider to switch kit suppliers as their enzymatic processes can differ and can result in other forms of data biases.

3.1.3   Pilot experiment

Many of the topics discussed before can be checked with the help of a pilot experiment. Aspects such as sample size calculation (how much variance is to be expected in the data?), severity of treatment effect, RNA quality and quantity need to be checked prior to the main experiment, especially if one is not already familiar with the experimental setting.

Pilot experiments are simple experiments that typically focus on a single variable using few samples. Difficulties that would arise in the main experiment – especially if the researchers intend to run many samples – can be easily discovered in a pilot experiment. This will help to save both time and money as it tends to be difficult to correct problems in the obtained data mathematically.

When one is not sure about the effect of a treatment on the sample, it is also very important to carry out a pilot experiment as many subsequent data processing steps prior to the analysis will make one or both of the following assumptions: (1) only a small percentage of genes are differentially expressed and (2) there is a balance between the number of up-regulated genes and the number of down-regulated genes. Especially when looking at changes in gene expression over time during development, differences between tissue types, or when using treatments causing a response of the immune system, these assumptions are likely to be violated.

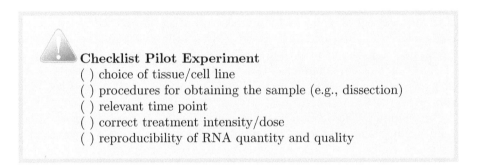

Checklist Pilot Experiment
( ) choice of tissue/cell line
( ) procedures for obtaining the sample (e.g., dissection)
( ) relevant time point
( ) correct treatment intensity/dose
( ) reproducibility of RNA quantity and quality

3.1.4 Main experiment

Once the main parameters of the biological experiment are defined, it is important to reconsider the biological question. This will define the strategy how one will analyse the data. Datasets from microarray experiments are huge and one can easily get lost in the many different ways of analyzing them. It is common that people who have little experience with running gene expression studies underestimate the time that is needed to analyze the data.

It can be very beneficial to try to answer one biological question at a time. Complex experimental designs with many different conditions (tissues/time points/doses/etc.) are usually also difficult (time consuming) to analyze. In this context it can also be beneficial to decide on one of two strategies: Is the experiment done to identify new transcripts/new functions or is one trying to learn something about the biology by focusing on genes with known function? Deciding one of the two will help in choosing the right arrays, e.g., exon arrays vs. normal chips. And it will also suggest whether the standard Affymetrix annotation can be used or whether one should prefer to analyse the data using alternative mappings between probes and probesets (alternative CDFs, see Section 2.2.2).

3.1.4.1 Control experiment

Is there any other experiment that can be done to confirm the successful conduction of the biological experiment? This can be an independent read-out such as a behavioral test/the determination of some blood chemistry/anything that is also expected to be affected by the biological experiment.

Having such an experiment that is done with the same treatment at the same time is highly important as it will help convince the experimentor of the relevance of novel findings as well as the accuracy of unexpected findings. As gene expression microarrays result in huge amounts of measurements, it is only one aspect to use statistical techniques to remove false positives, but it is another aspect to deal with the true positives.

As microarrays deliver so many measurements it is important to realize that statistics will only help in excluding false positives and identifying the truly changing genes. It is something else to deal with the true findings. The control experiment referred to in this section will help convince the experimentor of the relevance of novel findings or the accuracy of unexpected results. Such an experiment needs to be done with exactly the same treatment at the same time.

For example, when comparing the effect of different compounds on a cell line, it can happen that one compound suddenly gives much less signal than previous studies would suggest. Are the findings correct? Compounds show differences in how easily they are dissolvable and how stable they are in solution. It is possible that the concentration used in the experiment was too high and the compound precipitated without the experimentor noticing it. When running a functional assay that checks whether the same compound solution that was used for the microarray has in fact its characteristic property will ensure that the obtained data are correct. For example if an oncology compound inhibits the activity of a kinase which in turn will not modify its target protein by phosphorylation, one could use an antibody that detects this modification. By running a Western blot, one can quantify the amount of phosphorylated protein, thereby measuring the activity of the kinase. When the compound is applied one expects less signal in the Western blot as the compound will have inactivated the kinase. Such an experiment therefore assures that the compound solution applied for the microarray is indeed a solution with an active compound.

3.1.4.2 Treatment

How many genes are expected to be affected by the experimental treatment? This is an important question as subsequent normalization approaches rely on the assumption that only a limited number of gene (often less than 10%) are changing. If very many genes change and one would use a non-linear normalization, the effect would be that differentially expressed genes would appear to be less differentially expressed as they are. Experiments affecting global transcription or comparisons of cells from different organs are examples where one expects to violate the rule of limited gene activity changes.

3.1.4.3 Blocking

Whenever there is an element of the experimental design that will introduce technical variability into the data and that cannot be avoided, it is important to apply the concept of blocking. For example, if samples would need to be collected on multiple days, it would be essential to avoid collecting all control samples on day 1, all samples receiving treatment A on day 2, all samples receiving treatment B on day 3, and so on. In such a scenario, it would be impossible to conclude whether differences observed between treatments were actually caused by the treatment or by technical effects introduced by the fact that the samples were collected on different days. Statisticians refer to such a situation with the term "confounding." In such an experiment, potential treatment differences would be confounded with day-to-day differences. Therefore one should collect or treat samples from all groups on all days. In other words, all blocks (here all days) should contain all treatment groups.

3.1.4.4 Randomization

In contrast to blocking, randomization attempts to alleviate biases that are introduced by unknown and/or uncontrollable factors. Therefore it should be standard to randomize the samples for animal studies and also randomize or alternate the experimental treatment whenever possible. The main advantage of this additional step is that there are often technical shortcomings in the experimental protocol that one cannot overcome (see Section 3.1.2.6 as an example). Alternating or randomizing samples between treatment groups ensures that these uncontrolable sources of variability are equally probable in each experimental group. In other words, randomization is carried out to reduce bias.

3.1.4.5 Standardization

It is crucial that all steps of the sample handling starting with the biological experiment until the scanning of the microarrays are standardized. This also includes the storage and freezing conditions of tissues as well as the handling of the samples during the whole process.

3.1.4.6 Matched controls

If possible one should consider matching for confounding factors like gender, age, body mass index, etc. when processing human samples or samples with high biological variability. Looking at samples from diseased patients, it is often difficult to get good matched controls as the diseased patients will often have had some form of treatment the controls will not have had.

3.1.4.7   Sample size/replicates/costs

While there is no general rule defining the number of replicates needed per treatment group, one still needs to come up with a balance between the risk of unreliable results, the maximum of samples that can be handled within the biological experiment and the involved costs. In the following section we will try to provide you with some guidelines that can help in finding this balance.

In the past people used to assess the quality of a microarray experiment itself by running samples in replicates. The use of such technical replicates of the same biological sample has proven to be an expensive strategy and is at times also limited by the amount of RNA available from a sample. Advancements in the quality of commercial microarrays have made it possible to omit the control of technical variability[13]. Nowadays it is advisable to replace technical replicates with biological replicates as they generally show much larger variability than the variability introduced by the technical process.

Under controled conditions, when technical artifacts can be excluded as sources of extra variability, the status of the immune system of lab animals seems to contribute primarily to the variabilty seen in mice. Both genes for components of the immune system as well as genes encoding for products that mediate the immune response have been identified[59].

When it comes to the actual number of how many replicates are needed (we will avoid the statistician's answer of "the more the better"), we usually apply the following rule of thumb: three for cell line work, six for animal tissues and at least ten for human samples. Furthermore, when working with samples to classify them into two or more groups, one usually needs many more samples. If six are sufficient per treatment group to detect differential gene expression, one should have 20 per group for classification. On the other hand, when running a time course experiment, one can sometimes get away with fewer replicates per time point. See also Section 3.1.4.10 which looks at sample pooling as a possible strategy to reduce costs and Section 5.5.5 which discusses ways of calculating an appropriate sample size to detect an effect of a certain magnitude.

3.1.4.8   Balanced design

Even though there are numerous reasons why it is not always possible to come up with a balanced experimental design whereby each treatment group contains the same number of biological replicates, the data analysis of such balanced designs is much more straightforward.

However, especially when looking at bigger studies that tend to be designed in a balanced way, it is not unusual to end up with an unbalanced dataset.

---

[13]Reanalysing the MAQC data (see Section 7.2) Klebanov and Yakovlev quantitatively assessed the inherent measurement errors of the Affymetrix platform. They clearly highlight that "contrary to popular belief" the technical noise is quite low and has negligibly small effects on the results of statistical inference from the data[58].

Power describes the probability of correctly detecting that there is something happening (the null hypothesis is called false and it is indeed false).

StatsBox 3.1: Power

This can be caused by technical issues (e.g., chip failures, degraded RNA) or administrative problems, but also by experimental shortcomings such as toxic effects of treatments, higher drop-out rate, etc. that could not have been anticipated. Even though the analysis of an unbalanced dataset is more difficult, there are ways to deal with unbalanced data. It has to be noted though, that some simpler statistical tests that a non-expert might be tempted to use (e.g., a pairwise t-test) might no longer be valid for the data. Depending on the severity of the imbalance, such data will require support from a biostatistician to assist in applying the right analysis approach.

3.1.4.9 Control samples

What is the response when I do not apply a treatment? What happens to the system in the course of the experiment that is not caused by adding any treatment?

Be careful not to include samples that control for an effect that is better studied in a pilot experiment (see Section 3.1.3). For example, one is interested in the effect of a drug on a cell line. It is completely sufficient to have two groups: the compound treated samples and the solvent-only treated samples. Assessing the effect of the solvent on the cells does not need to be tested. This is better investigated in a pilot study if one would expect a substantial effect of the solvent on the cells.

This section refers to only the relevant controls for the microarray experiment. See also Section 3.1.4.1 which discusses control experiments to ensure the relevance of the microarray study and Section 3.1.3 on other aspects that should rather be studied in a pilot than during the real experiment.

3.1.4.10 Sample pooling

There are various reasons that will cause the experimentor to consider combining RNA from several samples and processing them together on a single microarray. The basic assumption of sample pooling is that the mRNA levels of a certain gene in the pool is very similar to the average expression from individual samples. Kendziorski et al.[60] have shown in a study using RAE230A arrays that indeed "most expression measurements from RNA pools are similar to averages of individuals that comprise the pool."

Cost   Often pooling is done to reduce the cost of the experiment as microarrays are still quite pricey (see Section 3.1.4.7). Especially when the cost of the sample or the cost of the preperation of the sample is much less than the cost of the microarrays, this strategy is applied as one can save money by needing only a smaller number of arrays for the experiment. However, if budget is the primary reason why one considers pooling, one should strongly consider whether it is possible to be more specific in the biological question that is being asked via the planned experiments. We have seen many experiments where people tried to achieve too many different objectives within the same experiment. For example, if one looks at the effect of a drug treatment on a cell line, it is not necessary to profile the cell line with and without the solvent in which the compound is dissolved, as this data will only give an answer about the effect of the solvent on the cells. Even though it can be scientifically interesting to know the impact of the solvent, it will not help in answering what effects are caused by the compound.

Insufficient RNA   Another situation where one can consider pooling is when the amount of RNA that can be obtained from a sample is insufficient for the standard labeling protocol. This often occurs when using techniques such as laser capture microdisection (LCM, see also Section 3.1.2.3).

Reference   A further application is the generation of large amounts of a stable reference. While this is routinely done for dual-color microarrays, an application for such a reference RNA when using Affymetrix arrays is to monitor the performance of the whole system over time to avoid deterioration of the data quality.

Between subjects variability   When pooling samples, one loses the information of the variability in the pool of samples. Many (if not most) statisticians will strongly advise against pooling. However, using pooled samples can also be done to intentionally reduce the variability between subjects.

Simplification of lab procedures   Pooling can be a strategy in certain situations were it can lead to simplifications in the lab procedures. Experiments that will not aim to describe results for an individual but rather for a group of subjects and therefore includes very many subjects, pooling can shorten processing times and avoids potential other artifacts like the generation of batch effects when having to split the labeling and hybridization over several days or weeks. While there is little advantage in pooling RNA from two or three samples per microarray in large studies, it has been shown to be advantageous to pool a larger number of subjects[60].

Disadvantages of sample pooling   The resulting data can contain a pooling bias. This term refers to the overall difference between the signal measured from the pooled sample and the mean of the individual samples that were used to construct the pooled sample. Shih et al.[61] have suggested that "a possible reason for the bias is that mixing of the RNA may cause some alteration of individual RNA contributions such that some samples dominate more than the others in the pooled expression." Furthermore, while this artifact will affect all the measurements, there seem to be a small group of genes which are specifically affected by pooling[62].

It has to be highlighted that samples that would have been considered to be technical outliers when studied individually (e.g., due to poor RNA quality), cannot be removed afterwards and will have an effect on the data obtained from the pool. This holds also true when including samples into a pool based on information that later turn out to be wrong. A typical example of this is tumor samples classified to belong to certain type based on histological findings that later on turn out to be originating from a different tumor type.

Furthermore, sample pooling will always result in a loss of information about the individual variability unless many pools were used. It is impossible to answer biological questions related to the individual. Applications such as the identification of subgroups in a disease via some form of clustering will not work as the pools will most likely contain members of different subgroups.

The worst case is certainly experiments that were to represent a treatment group by a single pool. As there is no way of estimating the variance between samples having received the same treatment, the only way of selecting genes is based on the observed fold change. Such a selection cannot incorporate any measure of how consistent or how reliable the observed change is. Small gene alterations will typically be difficult or impossible to detect. As all of the above-mentioned problems make it difficult to later on generalize the results of the study, one should consider results from experiment with single pools per treatment at best as screening exercises for further in-depth analysis, while keeping in mind the high likelihood of a number of false positive and false negative findings.

Alternatives to sample pooling A common alternative to pooling in situations of limited RNA amounts is the use of linear RNA amplification protocols. However, as they introduce biases into the data, they can sometimes not be the preferred choice, especially when one expects to see subtle gene changes.

3.1.4.11 Documentation

If the subsequent data analysis identifies outlier samples, it is of utmost importance to be able to trace back any issues that might have arisen in the course of the experiment. This again includes both the biological experiment as well as all the steps in the microarray experiment. Since it tends to be too difficult to document every single step of the process, the best way to be able to trace back mistakes is routine. Using the same way of pipetting, the same way of splitting samples during, e.g., centrifugation steps, array hybridization, etc. makes this possible. Furthermore, having randomized the order of samples initially also allows one to follow such a routine. Still, everything else that is out of the ordinary should be recorded. Examples include:

- Especially large or small tissue samples could correlate with either insufficient/degraded RNA or dissection artifacts due to removal of neighbouring tissue.

- A larger than normal air bubble inside the array cartridge can potentially result in lack of signal in the center of the array.

- Salt precipitations or comparably less amount of hybridization cocktail removed after hybridization could indicate leakage of the cartridge.

- Poor total RNA yields lead to diluted samples that need to be concentrated. This can also lead to a concentration of residual salts or other contaminants that inhibit the downstream reactions. It can often be observed that samples that had just enough total RNA yield still have less cRNA yields than other samples. These samples will often also result in poor hybridization and comparably less signal on the array.

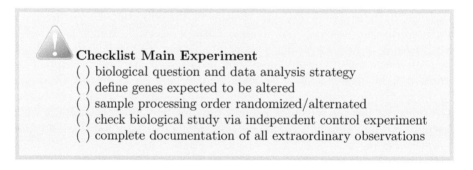

Checklist Main Experiment
( ) biological question and data analysis strategy
( ) define genes expected to be altered
( ) sample processing order randomized/alternated
( ) check biological study via independent control experiment
( ) complete documentation of all extraordinary observations

3.1.5   Follow-up experiments

Prior to the experiment one should also consider what one would do with scientifically interesting results. It is worthwhile to imagine a positive outcome and devise a strategy how one would confirm the findings. A classical approach is to use RTqPCR but it has been shown that there is an excellent correlation for the determination of fold differences between microarrays and other platforms for quantitative RNA measurements[15] such as TaqMan gene expression assays[63], standardized RTqPCR assays[64], QuantiGene assays[65], etc.

Depending on the confidence one has in the quality of the starting material, the number of samples, and the experimental design of the microarray study (the generalizability of the experimental results), one can also already take the

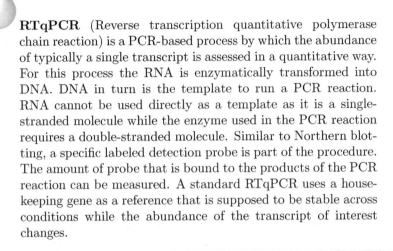

RTqPCR (Reverse transcription quantitative polymerase chain reaction) is a PCR-based process by which the abundance of typically a single transcript is assessed in a quantitative way. For this process the RNA is enzymatically transformed into DNA. DNA in turn is the template to run a PCR reaction. RNA cannot be used directly as a template as it is a single-stranded molecule while the enzyme used in the PCR reaction requires a double-stranded molecule. Similar to Northern blotting, a specific labeled detection probe is part of the procedure. The amount of probe that is bound to the products of the PCR reaction can be measured. A standard RTqPCR uses a house-keeping gene as a reference that is supposed to be stable across conditions while the abundance of the transcript of interest changes.

BioBox 3.8: RTqPCR

second step and confirm the findings in a potentially more informative setting: using, e.g., *in situ* histochemistry is still based on measuring the abundance of a transcript, but will also provide the scientist with the information where the transcript is localized in a sample.

As scientists are often not interested in studying the regulation of transcription, but typically rather like to use microarrays as a robust screening platform, one can also look into the availability of specific antibodies for a protein whose gene activity was affected in the gene expression experiment. Technologies such as ELISA or immunoblots can be used to confirm the observed changes in activity on the protein level.

3.2 Microarray experiment

3.2.1 External RNA controls

Commercial microarray platform providers such as the Affymetrix system offer a comprehensive panel of exogenous controls[66]. Two different types of "spike-in" controls can be used to check the quality of different steps of the assay process. They are either added to the RNA at the very beginning (poly-A controls *lys*, *phe*, *thr* and *dap*) or prior to the microarray hybridization (hybridization controls *bioB*, *bioC*, *bioD* and *cre*). The controls added to the hybridization cocktail are independent of the RNA sample preparation and are primarily intended to assess the efficiency of the hybridization. They are made of genes from the biotin synthesis pathway of *Escherichia coli* plus the recombinase gene from P1 bacteriophage.

On the other hand, the poly-A controls are spiked into RNA samples at the very beginning and are intended to identify issues occurring during the whole process (labeling, amplification, hybridization, washing, staining, scanning and data collection). Poly-A controls are genes from *Bacillus subtilis* that have an added poly-A tail, so that they can get reverse transcribed later.

3.2.2   Target synthesis

The target synthesis involves an amplification step to have sufficient material for the microarray. This step is not done by a PCR reaction which would involve an exponential amplification of the starting material. Such exponential amplification steps of complex mixtures (in our case a mixture of all RNA species that were present in a sample) have a limitation in how consistently the various RNA species are being amplified. As this usually leads to a distortion whereby some RNA species will be better amplified that others, the amplification step used in microarray analysis is a linear amplification which is thought to be equally efficient for all different RNA species.

Using exon arrays involves a different lab protocol whereby the method of generating the target that is hybridized to the array has been changed. While standard arrays make use of stretches of multiple A (adenosins) at the end as starting point for the generation of the target, the new protocol for the exon and tiling arrays uses random starting points. When the so-called "polyA-tail" of mRNA (a feature of most eukaryotic organisms) is used to generate labeled target, a bias is introduced into the data. This bias is caused by inefficiencies of the involved enzyme. Every time the enzyme does not generate a full length molecule covering the whole transcript, probes towards the starting point (the polyA-tail) will get a stronger signal than probes located more towards the end (the 5'-end of the transcript; see Figures 2.5 and 3.6). This effect has been taken into consideration when Affymetrix designed the standard arrays. Probes for these arrays are primarily selected to interrogate the 3'-end of a transcript. A positive side effect of this selection bias is that the labeled target can be reliably detected even if the synthesis of the target turns out to be inefficient as the T7 RNA polymerase starts the synthesis of biotin-labeled probe from the 3'-end.

The whole transcript sense target labeling assay used for exon arrays and tiling arrays is different to the standard protocol as it generates labeled DNA as target. The procedure for the standard arrays (e.g., the Human Genome U133 Plus 2.0) produces labeled RNA as target.

The better the integrity of the RNA the higher the yields from the cRNA synthesis. In other words, samples that yield relatively small amounts of

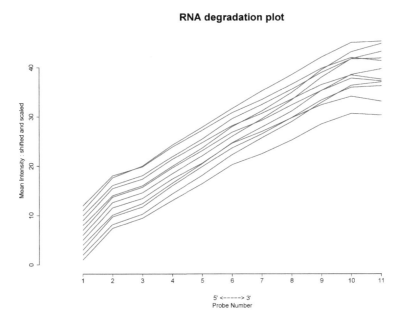

FIGURE 3.6: This graph is a plot to assess the quality of the RNA of the different samples used in an experiment. The observed average expression intensity across all perfect match probes of an array is compared to the relative position of a probe within a probeset. Typically, probes closer to the 5'-end show lower intensities than probes close to the 5'-end. This bias towards the 3'-end is introduced by the standard IVT labeling protocol. The closer a probe is to the 3'-end of a gene where the polyA tail is located on the mRNA, the stronger the signal intensity. Every blue line represents one microarray sample. If a sample contained degraded RNA the line of representing that sample will not run parallel to the other samples of the experiment and will have a steeper slope.

RNA due to poor quality will almost always result in poor cRNA and should therefore be excluded from the hybridization[67].

3.2.3   Batch effect

When isolating RNA and preparing labeled target material as well as when running the hybridization, washing and scanning, one should avoid processing samples on different days as this will introduce variability into the data[33]. Ideally, the maximum batch size that can be optimally processed should be established in a pilot experiment. Even though it is possible to remove such a day effect mathematically afterwards (see Section 4.2.5), such a step will also delete some of the biological signal. While this is less of an issue with biological experiments leading to strong differential expression, it is of concern when one expects subtle differences.

Besides such a day effect, there are many other potential causes for batch effects. A project could involve the processing of samples over a longer period of time whereby the RNA extraction and microarray processing are done in batches or even at different sites.

One other type of batch effect is caused by unintentionally mixing chips of different production batches that have large differences in their production date. Figure 3.7 shows a clear example. Here, a few chips used for a number of striatum samples came from a batch of chips that was two years older than the batch used for the other samples. While this is an extreme example and is clearly visible in the figure, the main issue was an overall decrease in signal intensity that was almost completely resolved by the normalization process. Still, such differences could dampen the biological variability sufficiently to result in a non-significant finding for a truly differentially expressed gene.

3.2.4   Whole genome vs. focused microarrays

When selecting focused arrays (see also Section 2.3.5) for a microarray one has to take extra caution either in the setup of the biological experiment or in the setup of the microarray experiment.

One purpose of focused arrays is to cut costs or reduce the amount of starting material as these chips tend to be smaller and contain only genes of interest for a certain biochemical topic or biological pathway such as toxicology-related genes or apoptosis. When using them, one expects large portions of the genes to be altered by the biological experiment. However, as soon as the experiment indeed results in a very high proportion of up-regulated genes or down-regulated genes, one has to be aware of the fact that the assumptions made during data normalization would be violated when using global normalization approaches such as quantile normalization (see Section 4.1.4).

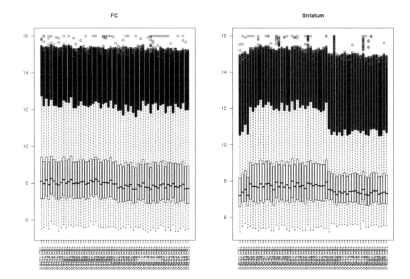

FIGURE 3.7: The two boxplots above show an example of a batch effect caused by using two batches of chips that differed largely by their production date. While the left boxplot showing frontal cortex (FC, a certain brain region) samples seems ok, the right boxplot of striatum (another brain region) shows clearly that there is a set of chips that have much smaller ranges of intensities.

Chapter 4

Data analysis preparation

Preprocessing microarray data aims at removing undesired sources of variation so that obtained expression estimates reflect the true changes (and non-changes) in mRNA abundance as accurately and precisely as possible.

4.1 Data preprocessing

It has been clearly shown that the right choice of a suitable pre-processing step is crucial to obtain reliable data from microarray experiments[68]. This is primarily due to the fact that raw microarray data of the different probes are noisy. Moreover, there are many different sources of variation that need to be corrected for. They originate at different stages during the whole process and include the manufacturing of the microarrays and all necessary enzymes and reagents (e.g., batch-to-batch variation), the biological experiment (e.g., RNA isolation, tissue heterogeneity, inter-individual variation) and the microarray experiment (e.g., reverse transcription of mRNA into cDNA, probe labeling efficiency). The data preprocessing step attempts to remove the technical sources of variation.

It is important to realize that a data preprocessing method is not generally valid, as it depends on certain assumptions that will not always hold given the large heterogeneity among microarray studies[69].

When talking about Affymetrix data preprocessing people usually refer to three main steps: background correction, normalization and summarization. These steps will be discussed in Sections 4.1.3, 4.1.4 and 4.1.5. Two other relevant steps that the user needs to be aware of are the calculation of the signal intensity for each probe and the log-transformation, which are discussed in the first two sections.

4.1.1 Probe intensity

The whole process of data preprocessing starts after the hybridized microarray has been washed and the lab computer has stored the image of the scanned microarray in a file with the file name extension .DAT. Since the images are

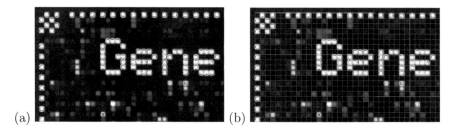

FIGURE 4.1: Checkerboard pattern in the corner of a scanned image. Figure (a) shows the pattern that is generated by spiking an artificial oligo into the hybridization cocktail which in turn hybridizes to complementary probes on the array. Figure (b) shows how the grid placement has identified the location of the different features.

saved in 16 bit-files where each image pixel can have one of $2^{\hat{}}16$ discrete levels of grey (65,536 levels), at most 4 orders of magnitude are possible to detect.

After image acquisition, Affymetrix' software (either GCOS or AGCC) will identify the position of every probe location (termed "feature") by placing a grid on top of the scanned image. The initial grid placement relies on anchor points in the corner of each array which are generated by probes against the B2 oligo (see the alternating signal along the border of the image as well as the checkerboard pattern shown in Figure 4.1). This B2 oligo contains a sequence that is unknown in currently sequenced genomes and can therefore be used to create a specific pattern using sequences that will either hybridize large amounts of the spiked artificial B2 oligo or detects no signal in the positions in between. To further improve the robustness of the extracted probe data (especially with arrays that have feature sizes below 18 micron), an extra step termed "feature extraction" is carried out by which the position of the grid is further adjusted for every feature.

After the grid is properly aligned on the image, and the location of each probe is identified, a square set of pixels is assigned to every probe (e.g., the size of the square for arrays that have a feature size of 11 micron is 5 x 5). From this set of pixels an overall signal value is determined as follows: the outer ring of pixels of the square is discarded as it is considered to be unreliable. Reasons for this limited quality are, e.g., some remaining misalignment of the positioned grid or the fact that a pixel in this outer ring could have picked up signal not just from the assigned probe but to some extent from the neighbouring probe as well. Of the remaining pixels, the pixel intensity of the 75th percentile is calculated. This respresents the overall probe cell intensity that is reported in the .CEL file. To provide the researcher with some guidance on how variable the pixel intensities for a given probe cell were, the standard deviation as well as the number of pixels that were used for the calculation of the 75th percentile are also reported in the .CEL file.

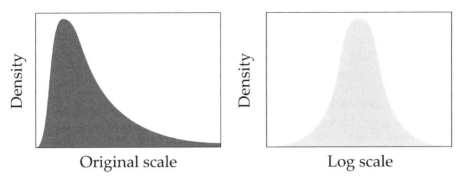

FIGURE 4.2: Data that is log-normally distributed in an original scale (left panel) and after log-transformation (right panel).

4.1.2 Log2 transformation

A typical first step in preprocessing microarray data is transforming the values to log scale, and applying all subsequent analysis steps on these log-transformed values.

There are several reasons why log-transformation is beneficial; two very important statistical arguments and one biologically interesting side effect. First, microarray intensities are typically asymmetrically distributed. Together with the low-expressed genes, there are many genes that are below the detection limit of the technology. As a result, there are many more probe intensities at the level of background intensity or just above the level of background and very few genes with very high expression levels. This makes it difficult to estimate certain characteristics of the intensity data and consequently of the expression data. For example, an estimate like "the mean intensity" will be inappropriate as it will be biased towards the few genes with very high intensities. Log-transforming the data makes the intensity distribution more symmetric and bell-shaped, i.e., a normal distribution (see Figure 4.2). Second, variation in intensities typically grows with the average intensities. This is known in statistics as the variance-mean dependency. The variation depends on the mean as larger intensities tend to be more variable. Consequently, the within-group variances are higher in those groups where the mean is higher. This is a violation of the general assumption of parametric models (see Stats-Box 5.13) that all groups should have similar variances. As log-transformation corrects for this mean-variance dependency, it increases the power of statistical tests. In other words, it increases the chance to detect true differential expression.

An interesting biological side effect is that log-transformation converts additive effects into multiplicative effects. The biological processes in cells, organs and whole individuals presumably act in a multiplicative way. Effects of amplification and inhibition in post-translational feedback mechanisms are probably multiplicative and not just additive. Log-transformation exactly makes the intensities and the expression levels behave in a multiplicative way. For example, for fold changes (see Section 5.5.2.1) a -1 on a log scale is equal to a 0.5 in the original scale. This implies that the expression has been halved. In contrary, a 1 on a log scale is equal to a 2 in the original scale and implies that the expression has been doubled.

When to do the log2 transformation? For example, when looking at the preprocessing algorithm RMA (see Section 4.1.5.2), the log-transformation step occurs after the background correction and the normalization has taken place and prior to the summarization step which is a median polish for RMA.

There are also other transformation formulas that may be helpful, like variance-stabilizing transformations[70]. Low intensities are often more variable in log scale due to background noise. The reason is that background noise, which is technical and not biological, is working in an additive and not in a multiplicative way. The log transformation is too 'strong' for this background noise, and reverses the previously discussed mean-variance dependency. A weaker transformation, like a cube root, could bring the variances closer, which attempts to equalize variation for all genes by combining an additive model (for the lower intensities) and a multiplicative model (for the higher intensities) within one transformation formula. However, while old generation preprocessing techniques like MAS 5.0 indeed had considerable variation at low intensities, more recent preprocessing techniques like (GC)-RMA or FARMS do not have this issue as they do a better job in background correction (see Figure 4.4). When one of the latter techniques is used, variance-stabilizing transformations are often not necessary.

4.1.3   Background correction

The background correction step aims to remove non-biological contributions ("background") to the measured signal. Typical examples are nonspecific signals such as unspecific binding of transcripts, background signal from incomplete washing of the microarray, background patterns across arrays, etc. This step has the largest impact on low abundance transcripts as small changes in the obtained signal have a large impact on the calculated fold change. Table 4.1 gives a small illustration of the effect when one adds or substracts differences of 5 to the obtained signal. This is done both for large and small signals and using a comparison between two samples A and B. While the impact of the background is negligible for the larger signal, it has a drastic influence on the weak signal.

One of the most commonly applied techniques for background adjustment is included in the Affymetrix software. The square area of the scanned microarray image is devided into 16 equally sized squares. For each square the 2nd precentile of all probesets within this square is declared to be the background intensity for this particular square. This approach is likely to be accurate as it is very unlikely that the 2nd percentile intensity would represent true biolog-

TABLE 4.1: Impact of slight differences in background on the accurate estimation of fold differences (change) between two samples (A : B) when looking at low- and high-abundance transcripts in two experiments (Exp).

Transcript	Exp.	raw (A : B)	change	log2 (A : B)	change
low abundance	1	20 : 40	2	4.32 : 5.32	1
	2	25 : 35	1.4	4.64 : 5.12	0.48
high abundance	1	2,000 : 4,000	2	10.966 : 11.966	1
	2	2,005 : 3,995	1.99	10.969 : 11.968	0.999

ical signal. Since there are numerous genes that are only active under certain circumstances (e.g., during embryonic development), the typical microarray experiment will detect transcription for only 40 to 60% of all the probesets on the array.

The actual background value that is subsequently substracted from a given probeset is equal to an average of the 16 squares whereby the distance of each of the squares from a particular probeset is considered by weighting. The further a square is, the lower the weight of the background in that square for the calculation of the weighted average for the probeset (see Figure 4.6).

When trying to remove the proportion of the signal that is due to unspecific binding to the oligos, one is faced with the problem that the extent of this phenomenon varies from probe to probe. Affymetrix has approached the problem of estimating the amount of unspecific binding by including MM probes for every PM probe (see Section 2.1). The GC-RMA summarization technique (see Section 4.1.5.1) tries a different approach by trying to predict the hybridization properties of a given probe based on its GC content[71].

Cross-hybridization (also referred to as "non-specific binding") is a phenomenon that adds to the above-mentioned unspecific signal. The term refers to the binding of labeled target sequences with a very similar sequence to the complementary oligonucleotide probes. For example, single nucleotide mismatches especially towards the ends of the probe, can still result in binding of some of these targets. If the stringency of the hybridization (primarily via a combination of temperature and appropriate salt concentration in the hybridization cocktail) and the washing process do not completely remove such incorrectly bound target sequences, an unspecific signal is detected.

4.1.4 Normalization

Normalization is a data pre-processing step by which one makes the different samples of an experiment comparable to one another. To this end, systematic differences between chips such as variation in overall brightness (the signal intensities of all probes for a given microarray) and spatial or print-tip effects are removed as much as possible. In other words, normalization aims to remove technical artifacts in the data and to retain the biological variation induced by the biological experiment. If an experiment would have been done perfectly, there would be no need to do a normalization step as there would be no systematic technical variation. It thereby becomes obvious that any normalization is closely linked to technical quality control. Identifying technical problems is necessary to properly correct for them during normalization.

However, such correction procedures will most likely remove some of the biological signal as well. The extent of how much biological signal is removed depends on characteristics of both the biological experiment and the technical quality of the microarray experiment. Since normalization techniques involve certain assumption on data properties within each sample and between samples, it is important to be aware of them. Most normalization procedures like quantile and loess normalization have three main assumptions:

1. The majority of genes is not differentially expressed.

2. There is symmetry in the expression levels between the up- and down-regulated genes.

3. The differential expression signal is independent of the average gene expression levels.

To assess whether some of these assumptions might be violated in a study, one needs to carefully consider the following questions.

- For the biological experiment:

1. Were the same number of genes detected in every sample group? For example, if one were to study developmental processes and one would look at samples from largely different time points like the brain of newborn mice vs. adult mice, it is quite likely that this will not be the case as there are a number of genes that are active only during development that are completely switched off once the developmental process is finished.

2. Were the experimental conditions chosen in such a way that only a small percentage of genes show differences between all of the samples? Were there any global shifts in transcription for the majority of genes as one would expect, for example when an experimental condition were to shut down parts of the transcription machinery of the cell? Many normalization techniques assume that this is the case (e.g., quantile normalization).

3. Are the number of up-regulated genes roughly the same as the number of down-regulated genes? This is usually difficult to answer when designing the experiment, but one has to be aware that this is also assumed for the commonly applied quantile normalization.

4. Are there approximately the same number of low-, medium-, and high-expressed genes present in each sample group? This does not mean that there may be no changes, but the data will be affected after normalization if one sample group (due to a biological treatment) were to have many more high-expressed genes as compared to the other sample groups.

5. Is the RNA content of a cell the same for each sample group? While the microarray procedure starts with the same amount of total RNA for each sample, the data from a biological treatment that were to impact the transcription process as a whole would certainly be inaccurate after normalization procedures such as quantile normalization. In essence this question is closely linked to the above question 2.

• For the microarray experiment:

1. Were there many defects on the microarray that need to be corrected for mathematically?

2. Was there a lot of variation in overall brightness between the microarrays?

The following paragraphs will give an overview of commonly applied normalization algorithms.

Linear scaling This approach (as implemented in GCOS / MAS 5.0) is also referred to as "global scaling." It calculates for each microarray a single factor by which each probeset is multiplied. The factor is chosen in such a way that after the multiplication each microarray has the same average intensity across all genes. Linear scaling procedure is done on the probeset level after summarization, so that every probeset is represented by a single number. However, there also implementations of this approach that work on the probe level (e.g., the "constant" scaling version as implemented in Bioconductor) prior to the summarization step.

Linear scaling may be advantageous in experimental designs in which researchers continously obtain samples as it avoids to renormalize all microarrays whenever a new sample has been analysed. There are, however, approaches emerging that are designed to deal with experimental settings that require a sequential addition of samples. For example, a new version of RMA (see Section 4.1.5.2), called "Reference RMA" (refRMA) builds an RMA model on the basis of a large set of biologically distinct Affymetrix microarray samples (> 1,500 samples of 144 different organ types). This pre-computed reference model is applied to the data of the microarrays of a new experiment[72].

GCOS/AGCC. Both GCOS (GeneChip Operating Software, former Affymetrix standard) as well as AGCC (Affymetrix GeneChip Command Console, current Affymetrix standard) are programs made by Affymetrix which control both the array washing stations and the scanner. The software is furthermore responsible for locating the probes on the scanned images, calculating intensities for each probe and generating summarized values for each probeset based on the MAS 5.0 algorithm by default.

Quantile normalization   This algorithm forces the distribution of probe intensities to be the same for all microarrays in the experiment. First an average distribution is calculated and afterwards the distribution of each microarray is adjusted to have a distribution identical to the precalculated average distribution. A downside of this approach is that a certain probeset will sometimes have exactly the same value across all microarrays.

There are different flavors of quantile normalization. For example, many Bioconductor packages usually make use of a sorting approach, while dChip fits a smoothing function. The effect of these different flavors can be seen when looking at the probeset intensities across samples. The sorting variant usually leads to probeset intensities that are exactly the same and the smoothing approach will give you slight differences between samples.

Variance stabilization normalization   This normalization approach, developed by Huber et al.[70], actually combines background correction and normalization. An assumption is that most probesets (less than half of all probesets on the array) are not differentially expressed. As the method is however quite robust, violations on this assumption do however not always lead to invalid results. In contrast to a simple log2 data transformation, which is often used in microarray data analysis (see Section 4.1.2), this procedure applies a *glog*-transformation (generalized logarithm).

The motivation for this normalization is that microarray data historically consisted out of two distinct noise components: additive noise and multiplicative noise (see Figure 4.3). While log-transformation corrected for the intensity level-dependency of the multiplicative noise, it inflated the additive noise component at lower intensities. VSN transforms the data in such a way that neither the multiplicative nor the additive component changes with intensity level. This additive-multiplicative noise is a great concern when background is insufficiently corrected or, as illustrated in Figure 4.4, when MAS 5.0 summarization is applied (as substracting the mismatches increased the variation

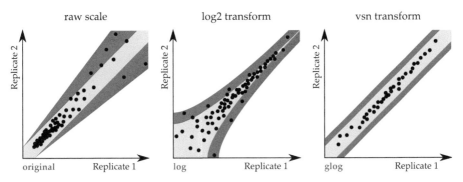

FIGURE 4.3: Schematic correlation plots of two replicates on raw scale (left panel), on log scale (center panel) and glog scale (right panel). As these are technical replicates, deviations from the equality line can be regarded as noise. There are two distinct types of noise components in microarray data: additive and multiplicative noise. The level of additive noise remains unchanged with increasing intensities at the raw scale (left panel), but gets inflated at low intensities after log-transformation (center panel). In contrast, multiplicative noise increases with increasing intensities (left panel), while log-transformation solves this mean-variance dependency (center panel). VSN normalization, which uses a glog-transformation, balances between these two approaches and stabilizes both the additive and the multiplicative variation. See text for more details.

particularly at lower intensities). When using the standard Affymetrix procedures like (GC-)RMA or FARMS, additive-multiplicative noise is rarely an issue (see Figure 4.4) making VSN transformation unnecessary.

An advantage of this normalization is that a "delta h" of 2, which is somewhat the equivalent of a two-fold change, will be independent of the observed intensity (see Figure 4.4).

Figure 4.5 shows that MAS 5.0 has more variation in log-ratios at low intensities compared to the VSN or RMA. This is *not* because the low intensity genes tend to be more differentially expressed, but because they just are more variable. This nicely illustrates the danger of using MAS 5.0 as it shows that MAS 5.0 presumably will generate a considerable number of false positive genes, i.e. some of the variable but low intensity genes.

Loess normalization This normalization technique has been traditionally used to remove print-tip related effects with cDNA arrays or to remove intensity-dependent dye-specific biases in dual color arrays. Again, the main assumptions are that the majority of genes are not differentially expressed and that there is symmetry in the expression levels of the up- and down-regulated genes. Furthermore, the differential expression signal should be independent of average gene expression levels and average to zero.

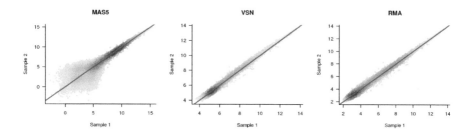

FIGURE 4.4: Comparison of intensity dependent variance between data normalized using MAS 5.0 (left graph), variance stabilization normalization (center graph) and robust multi-array average (right graph). Probesets with low intensities show higher variance than probesets with low intensities when using MAS 5.0.

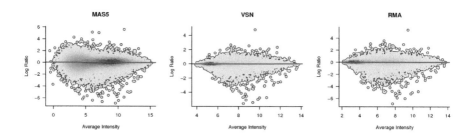

FIGURE 4.5: MA plot of differential expression by glucosamine treatment after IL1-beta challenge for three normalisation/summarization techniques. MAS 5.0 has more variation in log-ratios at low intensities compared to the VSN or RMA. The MA plot is discussed in Chapter 6, but basically plots differences in average intensities (M) on the y-axis vs. average intensities (A) on the x-axis for each gene (dot).

Loess calculations depend critically on the "span:" a value between 0 and 1 that specifies the range of data to include in each local estimate, and thus the degree of smoothing. The value specified for the span may significantly affect the results of loess normalization.

Rank-invariant normalization The main concept of this normalization approach is the following: if a gene is differentially expressed between two samples, it should have a higher rank in one array than another. Accordingly, genes that are not differentially expressed, should have similar ranks irrespective of signal intensities. For example, a gene that has the 450th highest signal intensity in sample A and the 455th highest intensity in sample B would be considered rank-invariant and would be used to arrive at the normalization factor. On the other hand, a gene that varies from 150th rank in sample A to 15,000th position in sample B in signal intensity would not be rank-invariant and would not be used.

Similar to the global scaling mentioned above, rank-invariant normalization uses a linear scaling of the arrays being compared. However, unlike the averaging method, the scaling factor is determined not by an average of all genes, but by these rank-invariant genes. Those "rank-invariant" genes show expression values with a consistent order relative to other genes in the population.

A major advantage of this normalization method is that it is much more robust against outliers than simple averaging. However, as with averaging, if a large percentage of genes varies between samples, the underlying assumption of rank invariance (the existence of a subpopulation of genes whose expression is constant across samples showing consistent ranks) will not be true and the method should not be applied. Furthermore, when there is a large difference in overall signal intensity between arrays, this normalization approach should not be used. Typically, rank-invariant normalization is fairly resistant to technical variations in large numbers of arrays.

No normalization If the experiment cannot be done in such a way that the normalization assumptions hold, it may not be considered to normalize at all. In this scenario one would need to conduct the experiment very carefully to avoid technical artifacts. One also should use technical replicates to remove any potential technical outliers and use log2 or probably even better glog-transformation. If done appropriately, clear effects will also be identified without normalization, although it is expected that the more subtle signals may be missed.

4.1.5   Summarization

Affymetrix arrays use multiple probes to measure the same transcript. One preprocessing step, the summarization step, consequently deals with summarizing these repeated measurements into a single value for a probeset.

This step is critical as the resulting single value should represent the transcript abundance as accurate and precise as possible. As different probes within the same probeset differ in their efficiency to hybridize to the same target[73] resulting in differences in signal intensities between the PM probes of the probeset, it is important how these differences are dealt with.

The resulting single expression value can also be distorted when probes of a given probeset are not measuring the right transcript. Among other reasons, this can be caused by the fact that the definition of where genes and their corresponding exons and introns are located is constantly evolving (see Section 2.2).

Most summarization techniques have a robust way of calculating the gene expression value for each corresponding probeset. Down-weighting probes with an outlying behaviour makes the obtained measures from microarrays with scratches or particles on the array surface still meaningful. The fact that probes of a probeset are scattered across the microarray ensures that this approach works for even larger artifacts on the surface as the probes that were affected were most likely coming from different probesets.

For the final data analysis and data interpretation, it is important to know that the composition of a top list of differentially expressed genes appears to be little influenced by the choice of background correction method and normalization strategy. Alternative probe mappings (see Section 2.2.2) as well as using a qualitative assessment (see Section 4.1.7) for deciding whether a gene was detected or not (the MAS 5.0 based absent/present call) also have only limited impact. Gene lists are however clearly influenced by the choice of summarization technique[74].

Different summarization techniques give different results. Some techniques do not use mismatch probes present on standard Affymetrix arrays (like RMA or dChip) and therefore tend to result in smaller fold differences between groups (higher accuracy at the cost of precision) while techniques such as MAS 5.0 subtract the mismatch intensity and get larger differences at the price of decreased precision (see StatsBox 4.1). Using spike-in experiments to judge the quality of summarization techniques is limited as the background does not vary. In real experiments, not only do the genes vary but also the amount of unspecific hybridization. This makes it more difficult to judge whether a certain technique is better than another.

4.1.5.1 PM and MM techniques

MAS 5.0 This is the default processing algorithm of the Affymetrix software that is used to operate the scanner and the washing stations when using the GCOS software. The MAS 5.0 algorithm (also referred to as the "Statistical Algorithm") uses a single-step method calculating a robust average signal for each probeset (Tukey biweight estimator). Background estimation is provided by a weighted average. The weighted average is calculated by dividing the microarray into 16 equally sized rectangular regions and select-

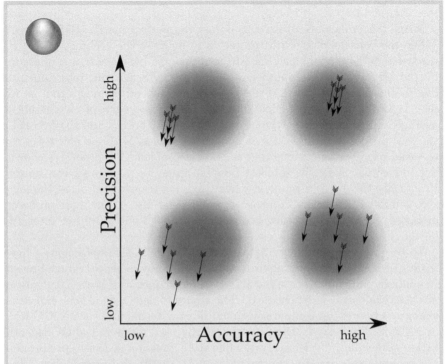

Accuracy and precision.

Accuracy reflects how close the measurements are to the actual (true) value.

Precision, also referred to as reproducibility or repeatability, reflects the similarity between repeated measurements of the same sample.

These concepts are often explained in analogy with firing arrows, where the repeated measurements are the arrows and the true value the target. When arrows strike closer to the target, they are considered more accurate. When arrows are grouped more tightly together, they are considered more precise.

Precision is often estimated as the standard deviation between the repeated measurements, or as the concordance correlation coefficient (but see Section 4.2.3.1). Precision is mostly estimated by the difference between the mean of the measurements and the reference value. This is also called the "bias."

StatsBox 4.1: Accuracy and precision

ing the lowest 2% of the feature intensities and averaging them in a weighted fashion depending on the distance of the 16 averages from a particular probe (see Figure 4.6).

Mismatch probes are utilized to adjust the perfect match (PM) intensity. While the concept of the MM probes is to increase the accuracy of the PM measurements by removing signal from unspecific hybridization as measured by the MM probes, the approach applied by MAS 5.0 runs into difficulties due to the fact that the MM probe picks up part of the real signal as well [9]. Learning from experience with the earlier approach of calculating an average difference between the average of the PM probes and the average of the MM probes resulting at times in negative values, MAS 5.0 now takes care to avoid negative values or other numerical problems for probesets showing low intensities. Now the smallest possible value for a probeset can be zero. Practically this poses a problem when using log-transformation as the log of zero is infinity. Therefore some people have either filtered out those probesets, or raised the intensities below one to one, or added one to all raw intensities on the microarray.

To be able to compare the data from all arrays in an experiment, a linear scaling of the feature level intensity values is used by default (trimmed mean). In contrast to many other summarization techniques, the MAS 5.0 algorithm analyzes each array independently. The ability to detect small gene expression changes is reduced in comparison to multi-array techniques such as GC-RMA or FARMS. Nowadays, MAS 5.0 is primarily used for technical QC (identification of technical outlier arrays), while the expression measures per probeset are typically calculated using one of the many multi-array techniques.

It is important to note that in contrast to many other summarization techniques, MAS 5.0 does the normalization step (the linear scaling using the trimmed mean) after summarizing the multiple measurements of the probes into a single probeset value.

GC-RMA GC-RMA[71] uses the MM probes to estimate GC-content specific hybridization characteristics. The effect of non-specific binding due the sequence of the probe are incorporated in the background adjustments. GC-RMA does not use all the individual MM probes for each PM probe.

GC-RMA essentially has the same normalization and summarization steps as RMA which is described below. The adjustment for GC content is only considered during the calculation of background correction. Like RMA, GC-RMA is a little less accurate but much more precise than MAS 5.0 (see StatsBox 4.1).

PLIER This summarization technique was introduced when Affymetrix started commercializing arrays with 11 micron features. A distinct feature of the probe logarithmic intensity error estimation (PLIER) algorithm is that it considers probe affinity data and target concentration to specifically detect

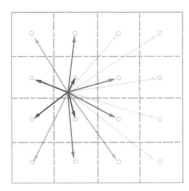

FIGURE 4.6: MAS 5.0 background calculation using a weighted average. The microarray is divided into 16 equally sized rectangles. An average background intensity is calculated for each region based on the lowest 2% of all intensities in the rectangle, indicated by green circles. For each probe, the distance is calculated to the center of every rectangular zone. A weighting factor is then calculated related to the distance. The colors of the arrows indicate the relative weights. Image courtesy of Affymetrix.

genes with low expression. Oligonucleotides show sequence-dependent differences in how efficiently they will bind to a complementary sequence. The probe affinity is trying to capture these differences by predicting a signal intensity for a given concentration of the target based on empirical data. Due to a scaling of the observed signal per probe, this multi-array approach allows to compare probes within a probeset even if the probes show absolute differences in intensities. In other words, the approach minimizes differences in affinity to the biotin-labeled target between the various probes.

The error model for PLIER considers both the target concentration as well as the background intensity. It assumes that the error is proportional to the probe intensity rather than the concentration of the target. Therefore, the error will be almost proportional to the target for high concentrations while it is proportional to the background for low concentrations of the target. As a result the higher degree of accuracy comes at the cost of increased signal variance.

For exon arrays a more advanced and efficient background correction method is used. This approach utilises surrogate mismatched probes. The signal variance is corrected by applying a variance-stabilizing data transformation. Small fold changes can be detected at the cost of increased variance.

4.1.5.2 PM only techniques

These techniques make use of the PM probes only. There are two major reasons why to discard the data of the MM probes:

1. The signals obtained from the PM probes are highly correlated with the signals of the corresponding MM values for the same transcript. This indicates that MM also detect the target transcript. This is in strong contrast with the purpose of MM probes, which were designed with an incomplementary base in the middle so that it would not hybridize the transcript[9].

2. For approximately one third of the probe pairs the signal intensities measured by MM probes are greater than those of the corresponding PM probes[9].

MBEI   The model based expression index (MBEI, also referred to as dChip or Li-Wong) was developed by Li and Wong based on their observation that the variation of a given probe across multiple arrays is considerably smaller than the variance across probes within a probeset[73]. Instead of using log-transformed probe intensity values, MBEI uses the original raw values as given by the .CEL-files. The algorithm takes probe-specific binding affinities into account by calculating expression values based on a model to estimate the signal based on the original scale. While MAS 5.0 attempts to correct for cross-hybridization signal by the use of mismatch probes, MBEI explicitly ignores these effects.

RMA   Robust multi-array average (RMA)[9] consists of three steps: convolution background correction, quantile normalization and summarization using a robust, multi-array linear model.

The background correction used in RMA is a non-linear correction based upon a convolution model and done on a per-chip basis. RMA assumes that all arrays have the same background. The correction step is based on the distribution of PM values amongst probes on an Affymetrix array. More specifically, the PM values are modelled as a mixture of (i) background signal intensities, caused by optical noise and non-specific binding, and (ii) an exponential signal component. During the background correction the intensities are forced to be positive.

RMA incorporates variation between probes in their affinity to the biotin-labeled target due to nucleotide composition variation. These probe-specific affinities are minimized by fitting a robust linear model to the observed probe log-transformed intensities across all samples of the experiment. By applying a quantile normalization at the probe level, the distribution of probe intensities becomes the same for all arrays.

After the probe-level adjustment, the value for a given probeset is summarized using median polish. Such a multi-array approach is advantageous when one is interested in detecting subtle changes between sample groups but is willing to accept some inaccuracies at the determination of the real fold difference.

sRMA This variant of RMA is optimized for data that are derived from samples with little amounts of RNA. To be able to run a microarray with such samples, one approach is to conduct an extra round of amplification. However, this introduces a bias towards the 3'-end. sRMA compensates for this artifact by weighting the probe intensities within a probeset according to their relative distance from the 3'-end[57].

FARMS Factor analysis for robust microarray summarization (FARMS)[75] also summarizes probe intensities using a robust, multi-array model. But while RMA uses a univariate linear model (see Section 5.5.2.5), FARMS is using a multivariate model (see Section 5.6.1 on page 198) per probeset. The intensity data of a single probeset is indeed multivariate as it contains many different probe measurements for many samples.

The core of FARMS is a factor analysis, a method to identify the "factors" that explain most variation in a multivariate dataset so that the many variables (here the probes) can be reduced into a few factors. Here only one factor is used, namely the first "common factor" that explains most of the variation within a probeset. This factor is then used to summarize the multiple probe intensities into a single expression value. The factor analysis is applied in a Bayesian framework (see StatsBox 4.2) so that important expected characteristics of the probesets can be incorporated in the model. One characteristic is that only positive relations between probe level intensities and probeset expression values are allowed. This is quite logical as increasing mRNA concentrations should only lead to higher, and not lower, signal intensities. Another characteristic is that most genes are expected to be non-informative, meaning that these probesets do not vary significantly across the different samples in the study. Particularly the introduction of this latter expected characteristic has an advantageous effect that is known as I/NI calls (see Section 5.3.1 on gene filtering). Like GC-RMA, it starts from log-transformed probe intensities.

FARMS has been shown to outperform most other methods both in sensitivity and specificity. In other words, it detects more signals while being more robust against measurement noise. Furthermore, it is also faster than the competitors[75].

4.1.6   All in one

Even though it is important to be aware of all the different steps which are done during the pre-processing of microarray data, oftentimes scientists are interested in complete solutions. The following will try to give guidance and selection of a suitable approach whereby we highlight the strengths of the various techniques.

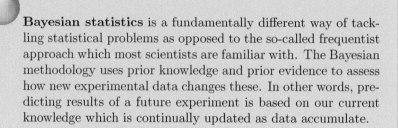

Bayesian statistics is a fundamentally different way of tackling statistical problems as opposed to the so-called frequentist approach which most scientists are familiar with. The Bayesian methodology uses prior knowledge and prior evidence to assess how new experimental data changes these. In other words, predicting results of a future experiment is based on our current knowledge which is continually updated as data accumulate.

StatsBox 4.2: Bayesian statistics

MAS 5.0

Implemented in Affymetrix GCOS and similarly in Bioconductor, using scaling to a target average intensity of 500.

- Single-array normalization will produce less normalization artifacts when the biological experiment contains treatments that resulted in very many gene changes (e.g., 30%) or in gene changes which primarily resulted in regulation in one direction.

GC-RMA

Implemented in Bioconductor, using quantile normalization.

- Multi-array technique leads to smaller variability of low-expressed genes.

FARMS

Implemented in a R package by the University of Linz, using no background correction, only PM data and quantile normalization.

- Multi-array technique which looks at consistency of probe behavior of all PMs in a probeset between samples of an experiment and thereby removes probesets that do not vary between samples.

4.1.7   Detection calls

This chapter explains different approaches to come up with a measurement that gives the scientist information whether the signal that was detected by the probeset was a true measurement of hybridized transcript or not. This type of measurement is often referred to as "detection call" or "absent/present call" and is commonly used to exclude probesets that were not detectable in any of the samples of an experiment.

4.1.7.1   MAS 5.0

Affymetrix has introduced one version of such a qualitative measurement in its microarray analysis suite version 5 (MAS 5.0). The rationale for doing this is that the MM probes give a reasonable estimate of background for the majority of the probes on a given chip, so if there is no statistical difference between PM and MM, then you might be able to consider that gene unexpressed. An indication for the accuracy of the MAS 5.0 detection call is described by Choe et al.[76] who showed that 85% of the true positives in their spike-in dataset were correctly classified as "present." Wilcoxon's rank test is used to calculate a significance or p-value and detection call for each probeset. Although the detection call is generated by MAS 5.0, this method can be used as a pre-filter to improve results using non-MAS 5.0 generated data, such as RMA.

4.1.7.2   DABG

With the introduction of arrays carrying PM probes only, Affymetrix had to replace its MAS 5.0 approach. DABG (detection above background) assesses the signal quality based on a comparison between the intensities of perfect match probes and probes that are considered to contain only background signal (see Section 2.3.2).

The process uses a collection of background probes (BGP) that have various GC content. The background correction is done using the median intensity of the backgound probes that have the same GC content as a given perfect match probe.

A major advantage of this approach over the traditional use of mismatch probes (see above) was the 50% reduction in the number of probes that are needed. This extra space can be used to roughly double the number of perfect match probes on the arrays.

For the detection call a score is calculated for each probeset. The relative distance between the perfect match probes of a given probeset and the background probes with the same GC content is calculated. Using this approach one can remove poor performing probesets prior to analysis.

4.1.7.3   PANP

Presence/absence calls with negative probesets (PANP), an algorithm written by Peter Warren, is based on the calculation of absent-present calls via the distance of a probeset to a collection of negative strand matching probesets (NSMP). This collection of probes was derived from the annotation of Affymetrix and represents probes that have no known hybridization partners. The NSMP probes are present on the array as the original design was based on EST matches with public databases. At the time of the design, a number of sequences were poorly annotated with respect to strand orientation, so that a set of probes are nowadays known to query the wrong genomic DNA strand (the reverse complement direction). Since NSMP sets have only been defined for the HGU133A and HGU133-Plus-2.0 microarrays, the application of this detection call is limited to these arrays.

PANP calculates a survivor function of the probability density of the NSMP expression values. Based on user-defined cutoffs, a detection call is generated which groups probesets into present, marginal and absent calls based on the interpolated intensities above, between, and below the cutoff intensities. Furthermore, a set of p-values is returned that indicate significance of the detection. Since the detection call is calculated on data after the summarization step, any preprocessing method (e.g., MAS 5.0 or GC-RMA) can be used.

The PANP detection call algorithm defines a threshold whereby probesets below the NSMP intensity are considered absent. In a few experiments we have come across a limited number of cases where based on I/NI-calls we could clearly detect biological signal below this threshold for the default absent calls. Therefore one should be cautious about using PANP for probeset filtering when the user-defined cutoffs are not carefully chosen.

4.1.8   Standardization

Standardization is the conversion process of values to dimensionless quantities. Standardization is the combination of two processes:

1. Centering: subtracting the population mean (or another measure for the center of the population) from each individual raw score.

2. Scaling: dividing the obtained difference by the population standard deviation (or another measure for the spread in the population).

In the context of microarray data, one may standardize genes and/or samples[77]. Samples are, however, rarely standardized as they are mostly already standardized as a kind of side effect of the normalization step (see Section 4.1.4). For the standardizing genes, the expression levels are transformed as follows

$$x'_{gi} = \frac{x_{gi} - center(x_{g.})}{scale(x_{g.})}$$

where x_{gi} is the expression value of gene g and sample i, and where $center(x_{g.})$ is an estimate of the center of the expression values across all samples for gene g (like the mean or median) and $center(x_{g.})$ is an estimate of the spread of the expression values across all samples for gene g (like the standard deviation or interquartile range (IQR))[77].

The result of standardization is that the different sets of values have similar distributions, and that the absolute values are conversed to relative values. For instance, genes can be standardized so that the average expression level across the samples will be zero for every gene. This removal of the "size component" can enhance the performance of some unsupervised projection methods as in spectral maps (see Section 5.4.3).

These dimensionless quantities are called "standard scores, z-scores or z-values." Standardization is sometimes also called "normalization," but in a microarray setting this definition should be avoided as normalization refers more to other preprocessing tools (see Section 4.1.4).

4.2 Quality control

4.2.1 Technical data

When looking at the technical data to decide whether the data from a microarray should be excluded from further analysis, especially the percentage of probesets called "present," the 3'/5' ratio of β-actin and GAPDH, as well as the scaling factor calculated by the Affymetrix GCOS software have been shown to be useful[67].

4.2.2 Pseudo images

Procedures such as the probe level based QC functions implemented in affyPLM[78] can be used to generate pseudo images that visualize potential artifacts or background gradients on the microarray surface based on plotting the residual weights (see Figure 4.7). These images show the residuals of a model fit based on probe level data. Using a logarithmic scale for the color coding of the residuals assures that only large residuals are highlighted and thereby the user's attention is drawn to more dominant technical problems.

4.2.3 Evaluating reproducibility

Comparison studies are often performed to assess the relative agreement in measurements of the same (or a highly similar) sample between two laboratories, two experiments or within one experiment.

In many -omics technologies, reproducibility is assessed by testing for a high correlation or concordance among replicates. As there are many measurements per sample, one can indeed easily check whether all pairs of genes of two replicates have relatively similar expression levels. This is however an over-optimistic and irrelevant assessment of reproducibility.

4.2.3.1 Measures for evaluating reproducibility

Correlation (coefficient) estimates the strength and direction of a linear relationship between two replicates (see StatsBox 4.3), and is mostly referred

FIGURE 4.7: Pseudo image generated with affyPLM which highlights a technical artifact in the middle. Residuals of a probe level model are color-coded whereby positive residuals are shown in red and negative residuals are shown in blue. Most likely this artifact was caused by leakage. A substantial amount of the hybridization cocktail leaked out of the cartridge resulting in a central bubble where labeled target could not hybridize to the probes.

Reproducibility and **repeatability** refer to the ability to accurately reproduce a replicated measurement or experiment. These definitions are often used interchangeably, although reproducibility typically indicates that the replicates are obtained more independently. Where reproducibility refers more to agreement between replicates across experiments (or laboratories), repeatability refers more to agreement between replicates within an experiment.

StatsBox 4.3: Reproducibility and repeatability

to as r. It ranges in values from -1.0 to 1.0. The closer the correlation coefficient is to either -1.0 (negative correlation or inverse correlation) or 1.0 (positive correlation), the stronger the correlation. A correlation coefficient approaching 0 indicates independence between the variables. While there are a number of correlation measures, the Pearson[1] and Spearman correlations are the most widely known.

Squaring the correlation coefficient gives a direct indicator of the relation between the variables. An r-value of 0.9 corresponds to a R^2-value of 0.81, meaning that 81% of the variation between the two variables is related.

 Correlation coefficients are inappropriate for assessing reproducibility. This is because the correlation coefficient is a flawed and incorrect estimate for agreement [81]. Replicates agree when their scatter lies along the equality line, but high correlation can be obtained if the scatter lies along *any* straight line.

Concordance (coefficient) estimates how close a relationship between two replicates lies along the equality line (i.e., the 45° line through the origin). Although this is generally regarded as an appropriate estimate of agreement *between two replicates using many genes* [81],[82],[83], it is still over-optimistic (see Section 4.2.3.2).

Intraclass correlation (coefficient) measures the relative consistency or conformity within multiple groups of replicates. More specifically, it estimates how close the replicate-to-replicate variation within samples is compared to the sample-to-sample variation [83],[84]. A value close to 1 implies that the sample-to-sample variation (the signal) is much larger than the replicate-to-replicate variation (the technical and/or biological noise). It is a useful and appropriate measure of agreement between *multiple replicates of multiple samples for a single gene*.

Unsupervised clustering measures multivariate similarities between samples. It is a very useful and flexible tool to assess agreement between *multiple replicates of multiple samples based on many genes*. There is a strong reproducibility when the replicates cluster together per sample.

---

[1] The theory of correlation is quite old. It was first introduced by Sir Francis Galton in 1885, and a decade later extended by Karl Pearson[79].

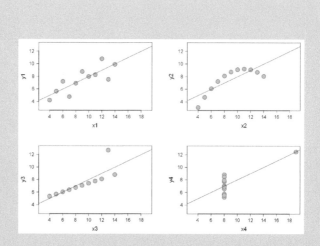

Correlation assumptions.
The picture above is an example of four different datasets. Besides having all the same correlation, they also have the same mean, variance and regression line[80].
Pearson correlation (see Section 5.4.2.1.1 for its calculation) is only appropriate for *linear* relationships of variables *without outliers*, as it is sensitive for outliers and assumes linear trends. It is therefore only appropriate in the upper-left situation and flawed in the other three situations.
Spearman correlation (see Section 5.4.2.1.1 for its calculation) is a non-parametric alternative which is more robust against outliers and therefore a preferred method for the situations visualized in the two lower panels. It is also better as a Pearson to estimate non-linear relationships as in the upper-right panel, but still not ideal. As this relationship is clearly quadratic, the preferred method would be to fit a quadratic regression curve.

StatsBox 4.4: Correlation assumptions

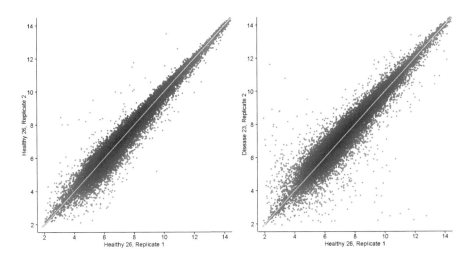

FIGURE 4.8: Panel (a) shows the correlation between two colon samples of the same healthy individual. Panel (b) shows the correlation between two completely independent colon samples: one from a healthy person and one from a diseased person.

4.2.3.2 A motivating example

Let's assess for example the reproducibility of gene expression measurements of human colon samples. In a study on irritable bowel syndrome (IBS), a gastro-intestinal disease, Aerssens et al.[85] obtained microarray data from colon samples of healthy and diseased people (see Section 7.3.1). Two colon biopsy specimens (10 cm apart) were collected from each participant, so that the agreement between biopsies of the same individual could be assessed. Figure 4.8a shows the correlation between two such replicates of a healthy individual. Although these replicates include besides technical also some biological within-tissue variation, they are strongly correlated (a Pearson correlation of 0.98 and a concordance of 0.98). However, a comparison between colon samples of a healthy person and a diseased person (Figure 4.8b) yields similarly high coefficients (a Pearson correlation of 0.96 and a concordance of 0.96). Note that there are some genes deviating from the equality line in Figure 4.8b; these are likely to be enriched with disease-related genes.

The reason why concordances between replicates for microarrays will always be high, is because microarrays measure everything. This includes some of the predominant, but uninteresting, sources of variation that are consistently high or low in all samples, like tissue-specific genes. A crucial housekeeping gene in colon tissue will be highly expressed in all samples, irrespective of disease status. A gene that is never expressed in colon, in contrast, will have low expression values in all samples. The ability to reproduce a big difference in expression level between a housekeeping gene and an unexpressed gene is meaningless. Such a difference is easy to detect, but is not the order of magnitude of the effects one generally is interested in.

Assessment of reproducibility should therefore not focus on within-individual reproducibility, but on its proportion vs. between-individual variability. This is exactly what an intraclass correlation measures (see Section 4.2.3.1). The disadvantage of this elegant measure is that it is a univariate gene-by-gene approach. However, exploring the distribution of the intraclass correlations across all genes might provide interesting results, especially if certain genes appear more reproducible than others.

Another solution is to center the data per gene so that all genes have an average expression level equal to zero. Reproducibility is consequently no longer assessed on the absolute expression values, but on the relative expression profiles. In other words, it estimates whether the replicates of a sample had similarly high or low intensities compared to the other samples in the dataset. Figure 4.9 shows the concordance correlations between the colon samples of the healthy and diseased individual used in the previous figure. It clearly illustrates that, without centering, the concordances are very high (coefficients > 0.95), even between samples that are expected to be quite dissimilar. The concordance after centering makes, however, much more sense. There is no concordance between the samples of the quite dissimilar individuals (coefficients around 0), and a significantly higher concordance between replicates of the same individual (coefficients of 0.55 and 0.57). Despite its attractive logic, using a concordance on centered data will only be interesting for large datasets without complex designs. In small datasets, or in datasets of complex design, the calculation of the residuals will be respectively imprecise or flawed.

Assessing reproducibility by intraclass correlation or by concordance on centered data is only possible if the experiment contains a relevant signal that one would like to detect. A weak intraclass correlation may have two potential causes: the method may be unreproducible (large noise) or the variation across samples may be weak (small signal). The experiment should therefore contain a relevant signal that one would like to find in the current or in future studies. This makes it practically more laborious compared to an assessment based on a comparison between two replicates. It is however worthwile the investment. Measurement error in absolute terms is actually not providing much relevant information. How measurement error relates to the signal(s) of interest is what matters.

The most informative way to assess reproducibility is by means of unsupervised exploration methods like spectral maps (see Section 5.4.3 for explanation):

1. As these exploration methods are multivariate by nature, they measure the agreement between samples using all genes.

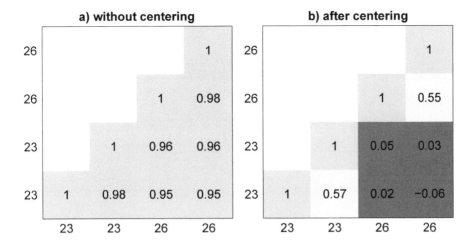

FIGURE 4.9: Pairwise concordance coefficients between four colon samples: two replicates from a healthy (nr. 26) and two from a diseased individual (nr. 23). Panel (a) shows their concordance without centering and panel (b) shows their concordance after centering.

2. Spectral maps project the largest sources of variation in a dataset onto a 2D figure. As the variation is made up by both replicate-to-replicate variation and the sample-to-sample variation, one can directly investigate their relative importance. The closer the replicates cluster together, the higher the reproducibility of the measurements compared to the overall variation.

3. There is no limitation on the number of replicates or the number of samples to include in the analysis. Each additional array will only imply an extra dot on the figure.

4. One can easily observe which samples are reproducible and which aren't. It is indeed quite common that, in a dataset with many samples, only one or a few samples have replicates that do not cluster together. This generally indicates that some of the replicates of these samples were precisely sampled, often due to dissection artifacts (see Section 3.1.2.6). An overall determination of reproducibility will fail to discover such a heterogeneity in reproducibility between samples.

Figure 4.10 shows the spectral map of colon samples of healthy and IBS-diseased individuals, each having two replicates. We now can show many more individuals than only the healthy and diseased individuals 23 and 26 used in the figures previously. The replicates of individuals 31, 37 and 44 (on the left of the figure) are quite reproducible, while other individuals like 16

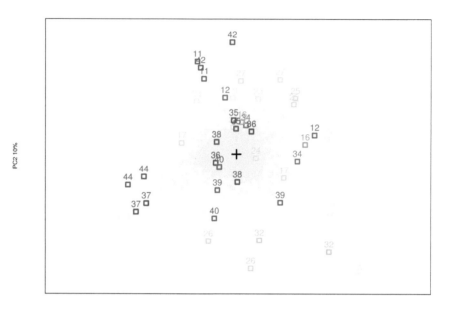

PC1 22%

FIGURE 4.10: Spectral map of colon samples of healthy and IBS-diseased individuals, each having two replicates. As each array is colored and labeled by individual, the two replicates of the same individual have the same color and label. The replicates of individuals 31, 37 and 44 (on the left of the figure) are quite reproducible, while other individuals like 16 and 17 (light- and dark-orange in the center) were less reproducible.

and 17 (light- and dark-orange in the center) were less reproducible. A closer look to the concordance of these samples can be used to decide whether their reproducibility is insufficient to include them in the analyses.

In general, there appears to be a good agreement between biopsies from the same individual as most replicates cluster together. The between-individual variation is higher compared to within-individual variation.

4.2.4 Batch effects

When a study is spread over time or when samples of an analysis were generated in multiple labs, it is important to assess how strong differences are between these batches of samples (see Section 3.2.3). But also well-controlled

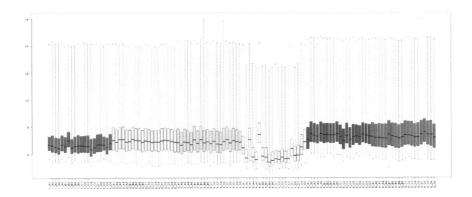

FIGURE 4.11: Boxplots of the Affymetrix data from the MAQC study before normalization. The boxplots of the samples are colored by their origin, i.e., one of the six test sites.

studies may still have been contaminated by an unforeseeable batch effect as examplified in Section 3.2.3 and in Figure 3.7.

It is therefore crucial to check for such non-biological sources of variation. Boxplots (see Figure 6.9) of the data before normalisation can help to discover batch effects (see Figure 3.7). Another useful and complementary tool are unsupervised methods like spectral maps (see Section 5.4 for more explanation on these methods). As these methods make use of the multivariate nature of the data, they can be very helpful to see obvious changes in gene expression profiles between samples. After observing a potential technical artifact, a search for the potential cause starts by examining if one of the technical variables, like experiment date or chip expiry date, can explain the observed patterns (see Figure 3.7).

Let's use the Affymetrix data from the MAQC study as a motivating example. Five replicate assays for each of the four sample types (A, B, C and D) were processed at each of the six test sites. As six test sites were involved, a logical first step is to check for batch effects due to test site.

Figure 4.11 shows the intensity values before normalisation for each sample in boxplots. The samples from site 4 (colored light-blue) have on average lower intensity values, except for three samples (A3, B1 and D5). The outlying behaviour of site 4 was also observed within the MAQC study. One of the potential explanations was that the technician from this site appeared to have very limited prior experience running the Affymetrix platform (Leming Shi, 2008, personal communication).

Figure 4.12 shows a spectral map of the expression values after normalisation and GC-RMA summarization. The x-axis of the spectral map, explaining 76% of the variation, clearly clusters the four sample types. The y-axis, however, indicates the presence of a clear site effect. All samples of site 6,

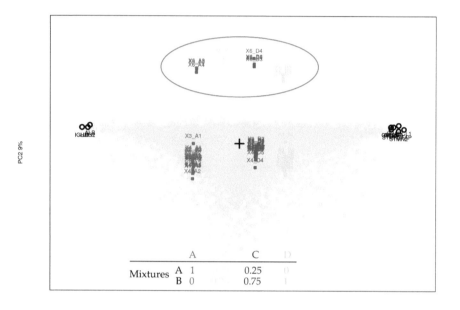

FIGURE 4.12: Spectral map of the Affymetrix data from the MAQC study. The 4 sample types A, B, C and D are colored red, lightblue, purple and green, respectively. The replicates of all sites are mixed within sample type, except for all the samples from site 6, highlighted by the orange ellipse. The legend at the bottom shows the ratio of RNA samples A and B used in the mixtures of the 4 sample types A, B, C and D.

highlighted by the orange ellipse, are discriminated from the samples of the five other sites. The difference between site 6 and the other five sites makes up 9% of the total variation in the microarray dataset. As the samples reflect extremely distinct RNA profiles, 9% is actually quite large. Also the MAQC participants noticed that site 6 showed very high intrasite consistency in sample profiles but significantly different from other test sites, but the underlying cause remains unclear (Leming Shi, 2008, personal communication).

   Interestingly, the relative configuration of the four sample types is identical for all sites. This indicates that the genes responsible for the batch effect of site 6 are not involved in the difference between the four sample types. This highlights a quite important practical consideration with respect to batch effects. If the batch effect is completely independent of the effects under study, it should be of no concern. In other words, genes that are not relevant for the study may differ between batches without harming the study. The problem is, however, that it is typically not known which genes are the relevant ones.

A final observation of Figure 4.12 is that the samples from site 4 cluster nicely together with the respective sample types from sites 1, 2, 3 and 5. In other words, the lower absolute intensity values observed in Figure 4.11 are not causing differences in relative values. This is a nice example for the need of normalization; the differences in absolute values were perfectly corrected so that the relative values were reflecting the true changes.

4.2.5 Batch effect correction

A batch effect is a source of non-biological variation due to an unfortunate experimental procedure in which the samples could not have been treated simultaneously, but were run in batches. Combining such samples without adjusting for batch effects might result in false negatives due to increased technical noise and/or in false positives when the variable of interest is confounded with batch. The latter implies that for example certain batches contained more treated samples and others more control samples. In the presence of a strong batch effect, certain genes might turn out to be significantly differentially expressed by treatment, while this is only due to the confounding batch effect.

The most logical way to correct for a batch effect is to model the batch in a linear model like a one-way ANOVA or LIMMA (see Sections 5.5.2.2 and 5.5.2.5) and to continue to work with the residuals of this model. Such an approach basically estimates the effect due to batch, and consequently adjusts the data for this estimated batch effect. At the end, each gene is centered to have a mean of zero within each batch.

A more elegant solution is to combine these two stages by modelling both the variables of interest together with the batch effect in one full model like a two-way ANOVA, also implemented in LIMMA.

Using linear models for batch correction can potentially destroy the correlation structure between genes. This is one of the most unfortunate side effects of using the residuals from a linear model for batch correction. As the correction is done gene-by-gene, and the estimated batch effects differ between genes, some gene correlations will become less obvious. Influencing the correlation structure can have a substantial effect on data analysis techniques that make use of this structure, such as certain pathway analysis algorithms. Furthermore, for batches containing few samples, the estimation of the batch effect may be quite imprecise.

An empirical Bayes approach to estimate batch effects in microarray data proposed by Johnson and Rabinovic[86] adresses most of the issues involved with using residuals for batch correction. This method assumes that phenomena resulting in batch effects will affect many genes in similar ways. By pooling information across genes in each batch, the individual gene estimates of the batch effect are shrunken towards the overall mean of batch effect estimates of all genes. As these empirical Bayes estimates are used to adjust the batch effects, the correction is done more robustly and the correlation between genes is less distorted.

After all analyses have been done, we generally plot the raw data of the top significant genes (y-axis) in function of the different batches (x-axis), with the dots colored by the variable of interest. For a top gene that was found to be significantly upregulated by treatment, it is important to check that its expression levels are on average higher in treated samples than in control samples in all or most batches. Such plots of selected genes relevant for the biological experiment will also reveal how strong the relative effect is within batches. Looking at potentially relevant genes and assessing the batch effect thereby helps in evaluating the evidence of that particular gene to be differentially expressed (see Section 4.13).

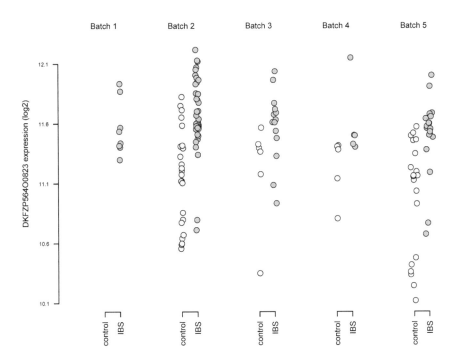

FIGURE 4.13: Intensity plot of a gene that is differentially expressed between IBS diseased patients and healthy controls in the IBS dataset. The data was shipped in five batches resulting in considerable batch effects. The first batch contained only samples from diseased patients. The gene DKFZP564O0823 has been shown to be differentially expressed after correcting for batch effects. This plot shows the uncorrected intensity values, and indicates that there is indeed a subtle upregulation in IBS patients within most of the batches.

Chapter 5

Data analysis

Simply generating gene expression data is clearly not enough. The generated data are only useful if one succeeds in extracting meaningful information from them about the system being studied[87]. Analyzing genome-wide expression data is a crucial task, as the choice of statistical technique has profound effects on the inferences drawn from the study. Appropriate and thorough analyses can discover so far undetected patterns in the data, while inferior approaches can mislead researchers and can spoil a study, by leaving interesting findings unnoticed or by suggesting falsely positive results.

Analyzing genome-wide expression data is a challenging task. First of all, there is no one-size-fits-all solution. Depending on the characteristics of the data and the goals of the experiment, different analysis techniques have to be chosen[87]. To choose the right statistic, one needs to understand to a certain extent both the biology and the concept of the study, as well as the advantages and pitfalls of the proposed analysis methods. Second, the popularity of transcriptional profiling and the new statistical challenges that accompany them has boosted the rise of many -omics analysis methods since the late 1990s, and this field keeps on growing exponentially. The available techniques are therefore numerous and often not fully matured, making the choice and the use of some techniques more difficult.

5.1 Why do we need statistics?

Statistical tests are sometimes seen as a necessary evil by researchers, who fear their complexity but know that they are needed to test hypotheses[88]. Researchers are often perplexed by statistics because of a language barrier between them and the statistician. This unnecessary but unfortunately common misunderstanding mainly exists because the conceptual purposes of the statistics remain unclear for the researcher. The most common cause for the existence of this language barrier is an excessive use of mathematical details and statistical terminology by the statistician. In contrast to the statistician, the scientist does not need to understand the underlying mathematical details, but should appreciate the rationale and the objectives of the statistics used. Conversely, the statistician may misunderstand the purposes of the study by a too technical description of the researcher. This has more drastic consequences as such misinterpretation may result in misspecified hypothesis formulations.

Here, we have tried to provide a comprehensive overview of the various steps in microarray data analysis together with their benefits and pitfalls. The more complex issues have not been circumvented, but are explained conceptually and intuitively.

5.1.1 The need for data interpretation

We need statistics for the interpretation of our experiments. The reason for this necessity is our wish to make generally valid conclusions although we cannot measure every individual. We therefore take only a random sample of individuals from the population of interest, and use this sample to make inferences about the population. If we for instance want to know which genes get differentially expressed in a specific disease, we will collect only a limited number of individuals, such as five diseased and five healthy ones. These obtained measurements are used to make conclusions on disease-related genes that are generally valid, and not just valid for these 10 studied individuals. To expand our conclusions from our sample to the entire population, we need statistics. This is because we will always find a difference simply by chance when performing experiments or measurements, even if we simply repeat a measurement or experiment. The key question is therefore whether the observed effect is real, or just due to the "inherent" variability of the experiments/observations.

There are many sources of noise and variability in microarray data, including experimental sources such as inconsistencies of image scanning, issues involved in computer interpretation and quantification of spots, hybridization variables (see also 5.1) such as temperature and time discrepancies between experiments, and experimental errors caused by differential probe labeling and efficacy of RNA extraction[89]. In addition, as the number of measurements increases, so does the probability of finding some large differences due to chance. Therefore, our conclusions bear uncertainty, and we need statistics to quantify this uncertainty so as to show that differences in gene expression are real.

The confidence that can be placed in conclusions drawn from samples depend in part on sample size. Small samples can namely be unrepresentative just by chance. As statistics also quantify the scope for chance errors, conclusions will become more significant with larger sample sizes even when the difference, for example between diseased and healthy persons, remains the same.

Parameters, variables, statistics. *Variables* define what you have measured. As it is impossible and undesirable to present all measurements, we want to describe the variables in a summarized way, for instance by averages and ranges. Such summarized values of variables are called parameters or statistics. A *parameter* is a number that is a property of the population and is mostly unknown, while a *statistic* is a number that is a property of the sample and is used to estimate the underlying parameter.

StatsBox 5.1: Parameters, variables, statistics

5.1.2 The need for a good experimental design

Statistics cannot replace a bad experimental design. Statistics extends our conclusions from sample to population level by taking random errors, like measurement error and uncontrolled variables, into account. Statistics can, however, not correct for systematic errors due to a biased measuring protocol or due to the collection of samples that are not representative of the population. Designing experiments is the work area of statisticians. To avoid irreparable damage, scientists should therefore ask for statistical advice before the experiment has started and not after the data has been collected[90]. Or, as R.A. Fisher said in 1938,

> "Consulting a statistician after an experiment is finished is asking for a post mortem examination. He can perhaps say what the experiment died of."
> Presidential Address to the First Indian Statistical Congress, 1938.

It is therefore crucial that, in an as early phase as possible, the lab biologist and the data analyst make a concerted effort to design experiments that *can be realized and analyzed*[5].

5.1.3 Statistics vs. bioinformatics

Comparing (bio)statistics with bioinformatics within the context of gene expression studies is a difficult discussion on semantics, especially because bioinformatics itself is a heterogeneous and loosely defined field. Statisticians and bioinformaticians mostly refer to one and the same thing in transcriptomics, namely to the process of interpreting the expression data. They, however, often approach the same problems from different angles, primarily because of separate historic evolutions that only started to assemble after the

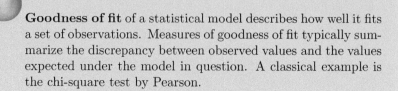

Goodness of fit of a statistical model describes how well it fits a set of observations. Measures of goodness of fit typically summarize the discrepancy between observed values and the values expected under the model in question. A classical example is the chi-square test by Pearson.

StatsBox 5.2: Goodness of fit

genomic revolution in the 1990s. The delay in evolving towards a unified approach in gene expression analysis is probably due to a language barrier between statisticians and bioinformaticians. This is quite unfortunate, as it is obviously clear that there is much to learn from one another.

Bioinformatics lessons for statisticians. Statisticians should learn more from the bioinformatical experience using large datasets. First, bioinformaticians usually have more experience with the development and implementation of tools to efficiently access and manage large datasets and different types of information. Second, the statistical community has almost exclusively made use of data models to reach conclusions on the collected samples[91]. Data models are well-defined types of models that incorporate some random noise, and can be validated using goodness-of-fit tests (see StatsBox 5.2). The other approach besides data modeling, called "algorithmic modeling" or "machine learning," has been developed rapidly in computer science. This rather recent culture of statistical modeling has shown to be of crucial importance when dealing with high-dimensional datasets because of its robustness against overfitting (see Section 5.6.2.1). This is because machine learning theories do not validate the model by goodness-of-fit criteria but by predictive accuracy criteria using independent test samples (see Section 5.6.2.3). In other words, a statistician will measure how well the model fits the data, while a machine learning scientist will look how accurately the model can predict the outcome of left-out samples.

Statistical lessons for bioinformaticians. On the other hand, bioinformatics can learn tremendously from statistics on how to interpret data. Statistics is an old research area; the modern statistics started with the introduction of the t-test, the chi-square and PCA in 1900-1908[92]. Being very active for more than a century, the discipline has accumulated numerous solutions to diverse problems with data interpretation. Statistics is now regularly seen as the guardian of good science because of its inextricable bounds with data interpretation. As pleaded by Vingron[5], the major upcoming challenge for the bioinformatics community is to adopt a more statistical way of thinking and to interact more closely with statisticians. A better statistical background knowledge will prevent some bioinformaticians to reinvent the wheel when developing seemingly new data analysis techniques, and a better statistical thinking will help bioinformaticians to use more "common sense" when conceiving and interpreting experiments. In cohort studies or more complex experiments there may be many hypotheses and potentially confounding factors, and statistical thinking helps to sift the wheat from the chaff.

5.2 The curse of high-dimensionality

The curse of dimensionality is the problem caused by the exponential increase in volume associated with adding extra dimensions to a (mathematical) space (Richard Bellman).

The curse of dimensionality is a significant obstacle in the solution of classification and clustering problems, as a high-dimensional feature space complicates the search for gene combinations with predictive power or the search for nearest neighbors (see Sections 5.4 and 5.6, respectively).

Another issue related to high-dimensional data is the problem of multiple testing (see Section 5.5.3). Testing a particular hypothesis gene-by-gene, for instance for a difference between two treatments, creates the problem of multiple testing. When looking at tens of thousands of genes, one will always detect transcripts which seem to behave in a desired way, unless the number of samples in the experiment is very large. These may be real, but may also be false positives that vary between treatments only due to random variability.

5.2.1 Analysis reproducibility

The reproducibility of microarray data analysis has been an issue over the past years[93]. This is often referred to by quoting a paper of Tibshirani and Efron[94] who tried to repeat the analysis of a well-known paper: "We re-analyzed the breast cancer data from van 't Veer et al.[95] [...] Even with some help of the authors, we were unable to exactly reproduce this analysis." While this sounds dramatic, the differences between both analyses were not that severe. Still, the fact that the findings could not be repeated exactly highlights the problem scientists need to be aware of.

The use of "point & click" software can turn out to be quite problematic in this context. Even though it is encouraged to look at the data with visual tools to increase the understanding of the study results, it is often difficult to document how a visual discovery came about and how to repeat it. A practical example is a software crash. Any discovery that happened before is difficult to reproduce and often requires that the user spends again a similar amount of time to re-identify a particular set of genes.

As microarray experiments regularly produce a large set of significant genes, another common problem occurs which is not always properly documented: the use of domain knowledge to select "interesting" genes because they seem to make sense to the scientist. While this is often fair to do and should increase the likelihood of focusing on the biologically relevant findings, it is strictly speaking subjective and the reasoning should always be documented.

Certain analysis methods, like some clustering and classification algorithms, make use of random numbers. Such methods will give slightly different results each time the analysis is re-run because the random number changes every time. To avoid this problem, many analysis tools allow the user to define this random number (often referred to as a "seed"). Using a specified number will ensure that the findings can be repeated.

5.2.2   Seek until you find

Similar to the multiple testing problem when selecting significant genes, the amount of data generated by microarrays causes another level of "multiple testing." Regularly, researchers do not stop after an experiment failed to answer the scientific question it was designed for. They continue their search and start incorporating other parameters in the analysis, or start new analyses that were not initially foreseen. Such exploratory analyses bear the risk to obtain false positive results, as the likelihood to find "something" significant increases the more tests you perform.

It is fair to maximize the knowledge that can be learned from a study, as microarray experiments take quite some time to do and are still fairly expensive. But the researcher needs to be aware of the problem of multiple testing and overfitting. To avoid this pitfall, he therefore should confirm the obtained findings in an independent experiment.

5.3   Gene filtering

Gene filtering is the procedure of removing genes that have no chance of being differentially expressed or predictive, regardless of the hypothesis or prediction problem that would be addressed. It is very fruitful as it increases the sensitivity of the analyses[96],[97], but it is still no common practice because clear guidelines are lacking and because it is rather dangerous as it bears the risk to exclude some potentially relevant genes.

Housekeeping genes are genes that are always expressed because they code for proteins that are constantly required by the cell. The proteins they code are generally involved in the basic functions necessary for the maintenance of the cell. Because they are essential to a cell, they are always expressed at similar levels under any condition.

BioBox 5.1: Housekeeping genes

Filtering vs. selection. Many people consider two different types of filtering, namely unsupervised or nonspecific filtering and supervised or specific filtering. Nonspecific filtering ignores the labels of the samples, while specific filtering uses the sample labels to select the genes. We strongly suggest not to use specific gene filtering, but rather gene selection to avoid possible confusion, as it is not a filtering process but rather a selection process oriented to find genes that are associated with a particular phenotype of interest. Examples are the use of fold changes (see Section 5.5.2.1) or significance levels (see Section 5.5.2.2) to select genes. We will discuss gene selection more in detail in Section 5.6 on classification.

The logic behind gene filtering is the fact that most genes are not relevant for the experiment. Microarrays can measure entire genomes at once. For a given experiment, focusing on a certain tissue in a few conditions, it is however a certainty that not the entire genome is of relevance. First of all, not all genes are expected to be expressed. Most tissues express only around 10,000-15,000 genes[98],[99]. Second, among the expressed genes, only a small fraction is expected to be differentially expressed by the different conditions used in the study[96]. Many genes are expressed for basic house-keeping purposes (see BioBox 5.1) and need to be expressed at a certain level for the maintenance of the cell. Most of these genes will consequently remain unaffected by any experimental condition.

The advantage of gene filtering is that it reduces the dimensionality of the dataset. The high number of genes on a microarray creates many practical and theoretical problems for data analysis (see Section 5.2). Much effort has been made to address these issues of overfitting and false positives due to multiple testing. These efforts have resulted in quite some breakthroughs like complexity-penalizing techniques to avoid overfitting and the false discovery rate (FDR). It is however very important to realize that these techniques are a cure, not a prevention. The FDR for example estimates how many false positives to expect, but it cannot identify which genes are actually the false positives.

In classification problems, correct and stringent filtering will substantially reduce the problem of overfitting. In tests for differential expression, filtering has two beneficial consequences: it will decrease the proportion of false positives in the top gene lists and it will diminish the impact of multiple testing corrections. The latter implies that the p-values of the true positive genes remain much more significant after the correcting for the number of tests applied.

Another interesting practical advantage of gene filtering is that it helps to solve problems with computer power or memory limitations, and will enhance the calculation performance of the computer.

5.3.1　Filtering approaches

There are several filtering methods of which some are often used in combination. They either start from the summarized data on probeset level (like intensity and variance filtering), or from the raw probe level data (like A/P and I/NI calls). The latter approach allows to incorporate knowledge on the behavior of the genes that is not been transferred in the summarization step.

5.3.1.1　Intensity of the signal

Filtering by signal (expression level) removes probesets with a signal close to background. Removing very low intensity genes makes sense because of the presumption that most genes have not been expressed.

Filtering by signal boils down to excluding genes that have an intensity level below the detection limit. Unfortunately, there is no clear detection limit in microarrays as it depends on the background levels that can vary from chip to chip. This makes the choice of intensity cut-off rather arbitrary. We personally often regard background intensities to have levels up to 3 or 5 on a $\log_2$ scale.

One needs to be sure to filter only the genes that are low in every sample. Otherwise one could filter out some of the most interesting genes, namely the ones that are expressed only under certain conditions. Regularly, one goes a little further and will exclude the genes that have low intensity signals in most of the samples (see Section 5.3.1.3 for more details).

5.3.1.2 Variation between samples

Filtering by the variation of the expression levels across samples removes the probesets that do not change in a given experiment. These probesets have a too low variability. But again the choice of what is "too low" is very arbitrary, creating the problem of which threshold needs to be chosen.

The variation in expression levels can be estimated in a way that is robust or sensitive to outliers. The inter-quartile range (IQR) is more robust compared to the standard deviation and will be less affected by a small number of extreme values. In complex experiments, one would like to retain genes that do not vary for most of the samples, but have outlying values in a few samples all belonging to the same condition. Here one should prefer a sensitive measure of variation like standard deviation. In a simple experiment containing some strange samples, one might want to choose a more robust measure as IQR, as otherwise genes would be retained simply because they have uninterestingly high variation because of these few outlying samples.

The IQR is calculated as the difference between the 75th and 25th percentile of the data. If one would use for example 2.5 as the arbitrary cut-off, one would claim that all the genes with an IQR smaller than 2.5 are not variable enough to be potentially interesting, and are consequently excluded for further analysis. See StatsBox 5.10 for how a standard deviation is being calculated.

The rationale behind variation filtering is to avoid the detection of genes of which the difference is statistically significant but too small to be biologically relevant. Indeed, the most common type of false positives obtained with t-tests is genes that are significant because their variation within the groups is very small but not because the difference between groups is large (see Section 5.5.2.2.1). Modified t-tests penalize these small variance genes (see Section 5.5.2.2), but the impact of multiple testing correction would of course decrease if these genes would have been excluded in the first place.

5.3.1.3 Absent/present calls

MAS 5.0 absent/present (A/P) calls is a measure that identifies whether the target transcript was detected or not by the probeset (see Section 4.1.7). A gene is called present in a certain sample when the PM probe intensities are statistically higher than the MM probe intensities by using a signed rank test [100]. A/P calls follow the same logic as filtering on intensity signal (see Section 5.3.1.1).

A/P calls can be used in different settings for gene filtering. The original procedure is to remove the genes that are called absent on all arrays. In other words, only the genes that are called present in at least one sample are retained for further analyses[100]. Indeed, a gene called absent in all arrays was in principle undetectable and therefore irrelevant for the study. McClintick and Edenberg[101] go a step further, and filter the genes based on A/P calls using more stringent proportions of a number of samples called present. Instead of "at least once," a gene needs to be called present in at least 50% of the samples in one of the treatment groups. If the study contains two conditions with each 6 replicates, then a gene needs to be called present at least 3 times in one of the two groups. The rationale is that a gene only detected once is still likely to be irrelevant if the different conditions in the study contain many replicates. If a gene is only being detected in a single condition, it should have been detected most of the time (i.e., more than 50%) to be a potentially relevant gene.

One needs to be careful with this approach, because the estimation of A/P calls is imprecise. It is known for quite some time that extra noise is being generated by subtracting the MM values from their PM partners ([9] but see [102]).

5.3.1.4　Informative/non-informative calls

On Affymetrix chips each target transcript is represented by multiple probes. The intensities of these probes are typically summarized for each probeset to provide one expression level for the respective target transcript[71]. Unfortunately, such summarization prevents the use of the information provided at the probe level. The idea behind informative/non-informative calls (I/NI calls) is to use this probe level information to determine how noisy the probeset is.

As the multiple probes are designed to measure parts of the same target transcript, they can be regarded as repeated measures of the same signal. This allows to estimate the reproducibility of a probeset. A highly reproducible measurement would imply that the probes measuring parts of the same target are strongly correlated. Now, one expects to observe only a reproducible signal when the array-to-array variation is similar for the repeated probes. When many probes reflect the same increase or decrease in mRNA concentration across arrays, their probeset is likely to be informative. In contrast, when most of the probes are not correlated, the change in intensity across arrays appears unreproducible across probes. The array-to-array variation is not exceeding the probe-to-probe variation within an array. This indicates that the probeset is non-informative, and could be excluded.

Figure 5.1 shows the first 5 probes of a non-informative (red) and an informative (green) probeset. In an informative probeset, the variation in mRNA concentration across arrays is apparent in all its probes, making these probes highly correlated. A non-informative probeset, on the other hand, has typically no consistent probe behavior. Here, increased expression values in certain arrays do not coincide in any of the joint probes. Empirical and simulated data show that probesets with an intermediate behavior between these two clear examples are called informative as soon as at least half of their probes are correlated[97].

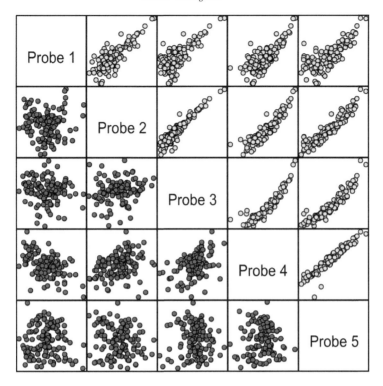

FIGURE 5.1: This scatterplot matrix shows the pairwise correlations among the first 5 probes of the same probeset across sample arrays for (a) an informative probeset (colored in green in the upper right panel) and for (b) a non-informative probeset (colored in red in the bottom left panel). Each dot represents an array.

5.3.2 Impact of filtering on testing and multiplicity correction

Gene selection, or specific filtering, cannot be used before testing for differential expression. It is based on a flawed logic as it actually runs two similar hypothesis tests after each other and uses only the latter for multiple testing correction.

The effects of non-specific filtering on testing and multiplicity correction, in contrast, is one of the most controversial issues currently in microarray analysis. Particularly whether or not to apply filtering before testing, and how to acknowledge for the filtering in the steps of testing and multiplicity correction is subject of much debate. One camp regards gene filtering as highly beneficial as it excludes the irrelevant genes, while the other camp advocates that gene filtering results in biased and overoptimistic statistics. This discussion is more of a statistical nature and is not addressing the question whether the removed

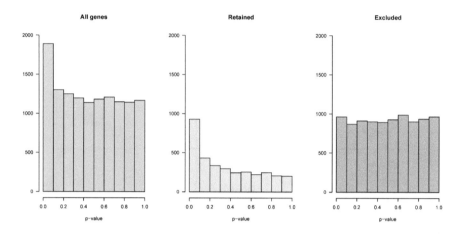

FIGURE 5.2: The distribution of p-values for the BCR/ABL-NEG comparison are shown for all genes (left panel), the retained genes after I/NI calls
(center panel) and the excluded genes by I/NI calls (right panel). The y-axis
shows the number of genes that are included in one interval. Note that the
sum of the retained and the excluded number of genes is equal to the total
number of genes.

genes indeed contain most of the non-informative genes and no informative
genes.

Let's illustrate the effects of gene filtering on the p-value distribution by
means of the comparison BCR/ABL vs. NEG in the ALL data. Figure 5.2
shows how the p-values are distributed for all the genes, and the retained
and the excluded genes by I/NI calls (right panel). The excluded genes (right
panel) follow a distribution that is expected if no gene is differentially expressed; due to chance, there are as many genes with low as with high p-
values. All the 1,000 genes with a p-value smaller than 0.1 can therefore be
expected to be false positive findings. In other words, I/NI calls indeed seem
to exclude non-informative genes. The distribution of the retained genes is
not uniformly distributed, but is enriched with genes with low p-values. In
other words, there are more significant genes than expected by chance. This
indicates that I/NI calls seem to include the informative genes, although there
are still genes retained that are non-significant.

Effect on testing. Gene filtering does not affect the raw, unadjusted, p-
values when applying ordinary t-tests. As these types of t-tests are done
gene-by-gene separately and do not borrow strength across genes for variance
estimation (see Section 5.5.2.2), each gene is tested completely independently.
Hence, the resulting p-value is unaltered irrespective of whether the gene is
analyzed together with few or many genes. Moderated t-statistics, in contrast,

make use of pooled variance estimates based on all genes in the data (see Section 5.5.2.2). Filtering on variance removes genes with small variance so that the estimate of pooled variance increases. But the effects are difficult to predict as changes in the used cut-off will also affect for instance the proportion of differentially vs. non-differentially expressed genes.

Effect on multiplicity correction. The effect of gene filtering on correction for multiple testing is straightforward. Gene filtering decreases the number of genes that need to be tested, and a smaller number of tests implies a less severe multiplicity correction (see Section 5.5.3). Hence, gene filtering results in smaller adjusted q-values, even if the raw p-value was unaffected by the filtering. With Bonferonni correction for example, the same p-value of 0.00001 results in 0.1 after correcting for 10,000 tests but in 0.01 when only 1,000 tests were performed. In other words, filtering can have a big impact on the absolute levels of the adjusted p-values. The ranking of the significance of the genes remains however unchanged. The top genes using the filtered data are exactly the same as the top genes of the unfiltered data, except the genes that were excluded due to having a too small variance or signal-to-noise ratio. Note that, although this varies a lot between studies, generally the proportion of such excluded genes in the top lists is smaller compared to the entire dataset.

Whether the multiplicity correction is valid after gene filtering is however not that straightforward. Quite some people regard gene filtering already as a type of hypothesis testing, and argue that the filtering step should not be ignored when adjusting for the number of tests performed. If ignored, the p-values would be insufficiently corrected for multiple testing, resulting in too small, overoptimistic, adjusted q-values. While this undeniably holds for supervised filtering (see above), it is not yet clear whether every type of non-specific filtering should be regarded as a real hypothesis test.

To filter or not to filter, that's the question. As with all statistics, there are trade-offs and assumptions one needs to make. We feel that a careful use of non-specific filtering has more advantages than disadvantages. Even if filtering might result in too small adjusted p-values, we expect this bias to be negligible. In any case, the p-values can still be used for ranking. Note that, at the end of the analysis, "low-quality" genes should be excluded anyhow from the top lists. Genes should be prioritized to decrease the possibility of moving forward with a false positive finding. A highly significant gene being inconsistent at the probe level or having a variance too small to be biologically meaningful should indeed not receive too much attention. Now, to enhance the efficiency and speed of our data analysis, we prefer to exclude them in the beginning rather than at the end.

Having said that, filtering bears a risk when implemented in automated analysis algorithms due to the inherent variability between different studies.

An option here is to use the filtering information only at the end. No genes are filtered out at the beginning, but the top lists at the end contain additional columns providing the information whether the gene was called informative and/or always absent.

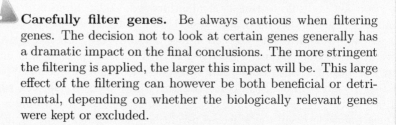

Carefully filter genes. Be always cautious when filtering genes. The decision not to look at certain genes generally has a dramatic impact on the final conclusions. The more stringent the filtering is applied, the larger this impact will be. This large effect of the filtering can however be both beneficial or detrimental, depending on whether the biologically relevant genes were kept or excluded.

Filtering of studies with a complex design. The *more complex the design* of the study, the *more dangerous filtering* becomes. This is basically due to the fact that a too stringent filtering can be flawed when only a small proportion of the samples is being differentially expressed. This warning applies to all types of filtering approaches (see Section 5.3.1). Let's illustrate this by means of two examples:

1. In a study with multiple groups, a relevant gene may be only differentially expressed in one group. For example, in a panel of 50 heterogeneous tumors, all done in triplicates, only one tumor may have a gene being over-expressed. This implies that only 3 out of the 150 samples change so that a too stringent filtering would prevent the discovery of this interesting finding.

2. In a study with many possible combinations of conditions, a relevant gene may be only differentially expressed in one specific combination of certain conditions. In a time series experiment with 5 time points after treatment together with a control, all having 6 biological replicates, genes may only be up-regulated by the treatment at the last time point and never by the control. This group of genes is therefore only changing in 6 samples of the 60 (6x5x2) samples, making them sensitive to be removed when the filtering criteria were too stringent.

5.3.3   Comparison of various filtering approaches

Let's compare the various filtering approaches using the 79 ALL tumor samples that were used to compare the BCR/ABL mutation with the NEG group. The samples were profiled using the Affymetrix Hgu95av2 chip, which

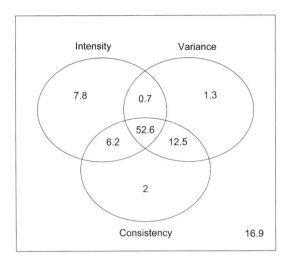

FIGURE 5.3: Venn diagram showing the percentages excluded genes for the three main filtering techniques: intensity, variance and consistency (I/NI calls). Note that there are 12,625 genes in total so that 10% corresponds with around 1,262 genes.

measures in total 12,625 probesets. We used I/NI calls[97] and the most popular versions of intensity and variance filtering[1],[7]. The used intensity filter retained genes having an expression level greater than $log_2(100)$ in at least 25% of the samples, and the used variance filter retained genes with an IQR larger than 0.5[1],[7].

Filtering based on intensity excluded 67.3% of the genes, variance filtering excluded 67.1% and I/NI calls 73.3%. The venn diagram in Figure 5.3 shows that 52.6% of the genes are excluded by all three filtering techniques. Most agreement is between variance filtering and I/NI calls; 65.1% of the genes (52.6+12.5) are excluded in both methods. This is expected as I/NI calls is an advanced form of variance filtering as it compares the array-to-array variation to the probe-to-probe variation.

The 2% genes excluded only with I/NI calls are generally genes with intensity or variation values close to the chosen cut-offs (the green circles in Figure 5.4). The intensity or variation of these genes was probably also too low, but just not low enough to fall under the specified cut-off. This again illustrates the issues that may arise with arbitrary cut-offs. The profile plots of two genes (genes 74 and 206) illustrate that their probes do not measure similar patterns across samples. Although having an intensity and variation exceeding the prespecified cut-off, these genes appear to have a random origin.

The combined filtering on intensity and variance would exclude 53.3% of the genes (52.6+0.7), so 0.7% genes that I/NI calls would have retained. In other words, these 93 genes have a low intensity and variation but their probes still detect patterns between arrays in a consistent manner. The latter indicates that, despite its low variation and intensity, the gene appears to reliably measure a potentially interesting phenomenon. Figure 5.4 indicates these genes with red circles on the top-left panel where intensity is plotted vs. variance. The profile plots of two genes (genes 14 and 17) show that the low average intensity and the low variation are mainly due to the fact that the genes are only expressed in one or a few samples.

The relation between standard deviation and IQR for all the genes is shown in the top-right panel. These two statistics, both of which estimate variation, are as expected to be almost identical in the absence of outliers. Most of the genes highlighted in red, however, have higher standard deviations compared to their IQR. This is because these genes are highly expressed in only a few samples (as illustrated in the bottom of Figure 5.4), and because standard deviations are much more sensitive to these outliers compared to IQR (see StatsBox 5.10 and Section 5.3.1.2).

That variance and intensity filters exclude the genes that are only expressed in few samples is very logical, as these filters have been deliberately chosen to do so. The typical filter for intensity is to require that genes must have an expression level greater than $log_2(100)$ in at least 25% of the samples[1],[7]. This results in the exclusion of genes that are highly expressed in for example 20% of the samples. The typical filter for variation is to require that genes must have an IQR larger than 0.5[1],[7]. As IQR is quite insensitive to outliers, this will lead to the exclusion of genes that have outlying expression values in a few samples as long as their overall variance is small. A choice for outlier-sensitive variation estimates like standard deviation, or a requirement that 100% of the samples need to have high expression levels, will lead to lower overall exclusion rates.

The main problem with filtering on variation and intensity is that it is linked to very subjective decision making as arbitrary cut-offs need to be chosen (see top-left panel of Figure 5.4). They do not seem to improve the filtering efficiency to a great extent. One could always increase exclusion rates by choosing more stringent cut-offs, but this would also increase the false negative rate – the exclusion of truly interesting genes. As it is quite crucial that no potentially interesting genes are excluded in microarray analyses, this false negative rate in the gene filtering step should be close or equal to zero. As there are no arbitrary choices attached to I/NI calls, it is much more objective and statistically more sound and less ad-hoc.

Figure 5.5 compares the average intensity, variation and significance of the retained genes between the different filtering methods. Intensity filtering retains primarily genes with average intensities higher than 5 and excludes most genes with average intensities lower than 7 (top panel). Variance filtering and I/NI calls do not have such a big impact on average intensities, but more on the variances; the retained genes are more variable (central panel). The p-value distribution (bottom panel) extends Figure 5.2 by showing the effects of variance and intensity filtering in addition to I/NI calls. The conclusions are

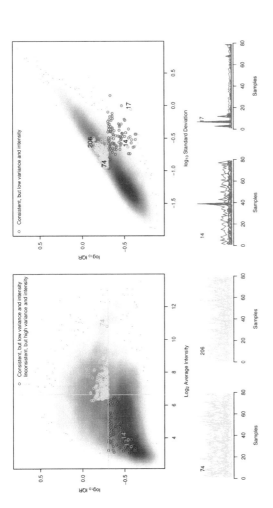

FIGURE 5.4: In the top-left panel, the variance (estimated by IQR) is plotted *vs.* the average intensity for all the genes. The green circles show the genes that are excluded with I/NI calls but not with the combined filtering of variance with intensity. The red circles show the genes that are excluded with the combined variance/intensity filtering but not with I/NI calls. The profile plots at the bottom show the profiles across samples for four genes: two that are excluded with I/NI calls and not with variance/intensity (the green profile plots of genes 74 and 206), and two from the other population (red profile plots of genes 14 and 17). The different lines on the profile plot show the behaviour of the different probes of the respective gene. It is clear from the red profiles that gene 14 is overexpressed in sample 39, as all its probes measure consistently this pattern. All probes of gene 17 indicate that it is overexpressed in samples 3, 7 and 12. In the top-right panel, standard deviation is plotted *vs.* the IQR. Most of the low variance genes that are excluded with the IQR filter deviate from the equality line because standard deviations are more sensitive to outliers compared to IQR.

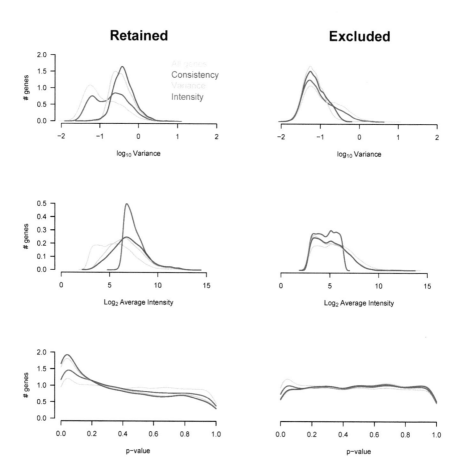

FIGURE 5.5: The distribution of average intensity, variation and p-values for the BCR/ABL-NEG comparison are shown for all genes (grey), the retained genes after I/NI calls (blue), variance filtering (red) and intensity filtering (green). The y-axis shows the number of genes that are included in one interval. Note that the grey density line showing all genes is used as a reference, and is evidently the same in the two panels of retained and excluded genes.

therefore the same; the excluded genes (right graph) follow a uniformal distribution expected due to chance, while the retained genes are enriched with genes with low p-values. Hence, all three filtering methods seem to succeed in retaining the informative genes. There is, however, a difference between the filter techniques in the degree of enrichment of low p-values in the retained genes. Intensity filtering results in a smaller enrichment, while I/NI calls result in the highest enrichment. Although this is no confirmatory evidence, this pattern suggests that I/NI calls are more successful in identifying the true positives.

5.4 Unsupervised data exploration

Unsupervised data analysis, or pattern discovery, is used to explore the internal structure or relationships in a dataset. Unsupervised refers to the fact that the labels of the chips are not used when analyzing the gene expression data (see StatsBox 5.3). These unsupervised learning techniques are typically multivariate, and can be subdivided in two main classes[103]: multivariate projection methods and clustering methods.

Unsupervised microarray data exploration already involves decision making, as we need to define what we believe is important. These decisions will drive the choice for a certain method and for a certain parameter setting within this method. For example, if we want to cluster the most similar samples together, then we can choose to apply a clustering method like hierarchical clustering. Consequently, we will need to define what we regard as being similar, and to define how similar samples need to be in order to assign them to the same cluster (see Section 5.4.2).

There are two important warnings with respect to the use of such unsupervised learning methods:

1. Unsupervised methods are an exploratory tool, and not an inferential technique. They are not appropriate as a visual inspection tool to identify differentially expressed genes. They do not fully incorporate the ratio of the signal (i.e., difference between groups) vs. the noise (variability within groups), and they do not consider the multiple testing problem. Having said that, an unsupervised method discriminating the groups of interest provides very strong evidence for differential expression between these groups. It suggests that the differential expression in certain genes between these groups is responsible for some of the largest sources of variation in the entire gene expression dataset. On the other hand, a failure to discriminate the groups of interest using unsupervised methods is not so meaningful. If only relatively few genes are differentially expressed, which is a common scenario, an unsupervised method

would fail to see differential expression, while a supervised method would succeed.

2. If a selection step (see StatBox 5.3) is applied prior to the clustering, for instance a selection of the most differentially expressed genes, it is wrong to interpret the resulting two clusters as biologically meaningful. Even a random dataset without any signal results in a nice clustering of the two groups of interest when based on genes selected as being the most differentially expressed. It is based on a circular reasoning as you will cluster what you have selected for and not the predominant source of information contained the data.

3. Unsupervised algorithms, particularly clustering methods, can be very unstable when learned on small datasets. This sometimes results in reproducibility issues[104],[105].

5.4.1   Motivation

Before blindly starting the analysis of the data, it is important that no obvious patterns in the data are being overlooked. This is why unsupervised data exploration is key, and should be the first step preceding other analysis approaches like testing for differential expression or classification.

Looking at the raw data is of primordial importance in data analysis. While statistics may deceive when applied inappropriately, the raw data are what they are and consequently cannot lie. When faced with only one gene, scatterplots of the expression values are therefore a simple but important tool because they visualize every single data point (see Section 6.1.2). Because of its high-dimensionality, microarray data can unfortunately not be inspected that easily. Only for looking at the raw data, we already need to rely on certain statistical methods to direct our attention to the more important features in the data. We need to define what we believe is important, so that we can choose certain methods and parameter settings based on this rationale.

5.4.1.1   Batch effects

Studies may be contaminated by an unforeseeable batch effect such as differences between samples that were collected at different time intervals (see Sections 3.2.3 and 4.2.4 for more information).

A spectral map is a very valuable tool to check for such non-biological sources of variation. A motivating example has been given by means of Figure 4.12, showing a spectral map of the Affymetrix data from the MAQC study.

5.4.1.2   Technical or biological outliers

Microarray datasets regularly contain samples with an outlying behavior compared to the other samples of the experiment. This can be due to technical

Microarray data algorithms are generally categorized into supervised or unsupervised methods.

- **Supervised** methods use the presence of the outcome variable (often called the "labels" of the samples) to guide the learning process. There are two types of supervised models:

    - *Classification models*: many genes at once used as explanatory variables (see Section 5.6)

    - *Tests for differential expression*: one gene at a time used as response variable (see Section 5.5)

    Section 5.6.1 contains a more detailed discussion on the conceptual differences between these two supervised approaches.

- **Unsupervised** learning methods only use the observations of the features (i.e., the observed expression levels of the genes) with the samples being unlabeled. A typical example of an unsupervised method is clustering (see this Section). These methods are used to group a (large) number of samples into as many classes as seem appropriate depending on the data (class discovery).

There are also so-called **semi-supervised** methods that are generally two-step approaches with one step being supervised and the other unsupervised. Examples are supervised PCA[106] and gene shaving[107].

StatsBox 5.3: Supervised vs. unsupervised learning

reasons, as subtle technical outliers can still pass a thorough quality check. Such samples should be discarded for further analysis. To identify whether the cause of the outlying behavior is of technical nature, one needs to go back to the quality check parameters described in Section 4.2. A remarkably low mRNA yield, or a low 3'/5' ratio of an Affymetrix control gene, for instance, may indicate that the sample was an outlier due to technical reasons.

No apparent deviations in technical quality measures suggest that the sample is biologically dissimilar from the other samples. One therefore should focus on all potential parameters that might explain why this sample is so remarkably different. A clear example from our own experience was an outlier in a gene expression profiling experiment of various tumor cell lines (data not shown) that appeared to be a tumor of a monkey while other samples were derived from human tumors.

5.4.1.3   Quality check of phenotypic data

Applying unsupervised methods to data of a study with a clearly defined hypothesis might seem counter-intuitive. As the hypothesis was defined prior to data collection and as the relevant information is available on each sample, supervised analysis is the most logical way to proceed. However, an important consideration is the quality of this relevant information, as this information might be incomplete[104].

Take for example the search for a gene signature for an heterogeneous disease like cancer or depression. Even after rigorously selecting patients, there may still be patients in the trial that are diseased because of a different cause, for instance due to different mutations in the case of cancer. Such a "bad quality" of the diagnosis inevitably results in suboptimal supervised data analysis. An unsupervised analysis may, however, reveal the presence of these disease subtypes. This is quite crucial as it may eventually lead to diagnosis refinement based on molecular evidence.

5.4.1.4   Identification of co-regulated genes

An unsupervised analysis of the expression levels of all genes in a microarray study reveals which genes behave similarly across the different samples in the study. It is consequently very useful as a first exploratory step in the search for co-regulated genes.

5.4.2   Clustering

Cluster analysis is an unsupervised technique by which either samples or genes or both are grouped based on the pairwise similarities between the samples/genes and certain distances between groups.

5.4.2.1 Distance and linkage

There are two important choices generally involved in clustering methods, namely the choice how the distance and the linkage need to be measured. The distance measure will determine how the similarity of two elements is calculated, and the linkage will define how similar elements need to be in order to assign them to the same cluster.

5.4.2.1.1 Distance measures for clustering

The distance measure, often also called "pairwise similarity" or "dissimilarity measure," results in a similarity matrix, a symmetrical matrix of scores expressing the similarity between two elements.

Euclidean distance. This is one of the most commonly used distance measures. Using Euclidean distance, genes/samples with similar absolute expression levels will be clustered together.

$$d_E(x, y) = \sqrt{\sum_{i=1}^{n} (x_i - y_i)^2}$$

The left part of the equation refers to the euclidean distance d_E between elements x and y. These elements can be either genes or samples depending on the research question. The right part boils down to the sum of the squared differences between the two elements over all the n measurements. If the elements are genes it is the sum over all n samples, if the elements are samples it is the sum over all n genes.

Manhattan distance. This measure is also referred to as city-block distance. In essence it is a robust version of the Euclidean distance, as the effect of single large differences (outliers) is dampened (since they are not squared).

$$d_M(x, y) = \sum_{i=1}^{n} |x_i - y_i|$$

Pearson correlation. Using the Pearson correlation, genes/samples with similar profiles across samples/genes will be clustered together. This distance measure is more sensitive to the shape of the gene expression profile than to the absolute expression level.

$$d_P(x, y) = 1 - r_{xy}$$

$$\text{with } r_{xy} = \sum_{i=1}^{n} \frac{\sum_{i=1}^{n} (x_i - \bar{x})(y_i - \bar{y})}{\sqrt{\sum_{i=1}^{n} (x_i - \bar{x})^2 \sum_{i=1}^{n} (y_i - \bar{y})^2}}$$

Spearman correlation. The Spearman correlation is a special case of the Pearson correlation. The coefficient is calculated similarly as the Pearson correlation, but the values of x and y are first converted to their respective ranks. Because it is not based on absolute quantitative values but on relative ranks, this non-parametric alternative is more robust against outliers than the parametric Pearson correlation.

5.4.2.1.2   Linkage functions for clustering

Single linkage (nearest neighbor).   The linking distance, or cluster-to-cluster distance, is the distance of the two closest objects between two clusters.

$$D_{AB} = min(d(u_i, v_j))$$

where $u \in A$ and $v \in B$ for all $i = 1$ to N_A and $j = 1$ to N_B.

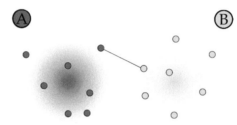

Complete linkage (furthest neighbor).   The linking distance is the greatest distance between any two objects in the different clusters.

$$D_{AB} = max(d(u_i, v_j))$$

where $u \in A$ and $v \in B$ for all $i = 1$ to N_A and $j = 1$ to N_B.

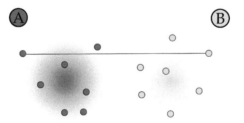

Average linkage. The linking distance is the average of all pair-wise distances between members of the two clusters.

$$D_{AB} = \frac{\sum \sum (d(u_i, v_j))}{N_A N_B}$$

where $u \in A$ and $v \in B$ for all $i = 1$ to N_A and $j = 1$ to N_B.

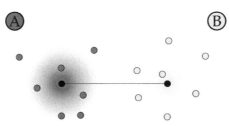

Ward's method. This method is distinct from all other methods because it uses an analysis of variance approach to evaluate the distances between clusters. In short, this method attempts to minimize the sum of squares (SS) within and maximize the SS between any two (hypothetical) clusters that can be formed at each step[108]. In general, this method is regarded as very efficient, although it tends to create clusters of small size.

5.4.2.1.3   Which distance or linkage to use?   There are many ways to measure distances and define linkages, and their choice has a big impact on the outcome of a clustering. Let's use a hypothetical example to illustrate this impact. Expression levels of 20 genes are generated for 10 samples belonging to two groups (A and B). The genes are simulated so that there are 4 gene families with distinct profiles, each containing 5 genes:

1. Profile 1: genes with relatively high expression levels and being differentially expressed (up-regulated in group B)

2. Profile 2: genes with relatively high expression levels and no differential expression

3. Profile 3: genes with relatively low expression levels and being differentially expressed (up-regulated in group B)

4. Profile 4: genes with relatively low expression levels and no differential expression

Figure 5.6 shows the striking effect of choice of distance measure on clustering performance. Based on euclidean distance, the genes with similar absolute expression levels are clustered together (profiles 1 with 2, and 3 with

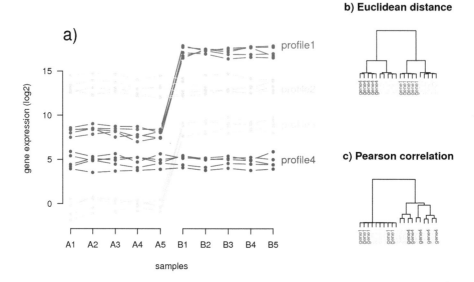

FIGURE 5.6: Simulated gene profiles to illustrate the differences between Euclidean and Pearson distances. Panel (a) shows the data; four types of profiles, each containing 5 genes, are being simulated so that they have a particular trend across the 10 samples (two groups A and B with 5 samples each). (b) Based on Euclidean distance, the genes with similar absolute expression levels are clustered together (profiles 1 with 2, and 3 with 4). (c) Based on Pearson correlation, the genes with similar relative trends in expression levels are clustered together (profiles 1 with 3, and 2 with 4).

4). Based on Pearson correlation, the genes with similar relative trends in expression levels are clustered together (profiles 1 with 3, and 2 with 4).

In a typical microarray experiment, samples are collected in different conditions to allow the study of the effects of these conditions on gene expression. In this scenario, researchers are primarily interested in genes with expression profiles changing over the different samples, and not in genes that are constantly high or low expressed. When clustering genes, there is generally more interest in clustering genes with similar changes in expression levels across conditions, irrespective of whether their average expression level is high or low. Under this rationale, Pearson correlation would be the preferred distance measure. An analogue type of reasoning can be followed when clustering samples. The majority of the genes will not change over samples, and clustering based on absolute expression levels is consequently less appropriate. It is more interesting to see clusters of samples with similar changes in expression profiles, for instance due to a group of 50 genes that are over-expressed compared to other samples.

As pointed out by Gentleman et al.[77], the behavior of the distance is closely related to the scale of the measurements. Whether the data were standardized is consequently an important issue when choosing the distance measure. If the data are standardized, it is more logical to apply Euclidean (or Manhattan) distances, as one is already looking at trends. This is because standardization has already converted the originally absolute values into relative values (see Section 4.1.8).

Given the fact that different distance measures highlight different biological patterns, the selection for a particular distance method should be primarily based on biological considerations rather than statistical ones. In general, however, we agree with D'haeseleer[109], who recommends the use of Pearson correlation as distance measure when using the raw absolute gene expression levels. In contrast to D'haeseleer[109], we prefer to cluster with Ward's linkage and not with complete linkage. We prefer to work on the absolute rather than the relative values, as standardization bears the risk of removing some potentially interesting features in the data[77].

Real-life example. Let's use the ALL data as a motivating real-life example. The ALL data contains samples of two types of ALL, namely B-cell and T-cell. Although both are lymphocytes, B-cells and T-cells have different roles in the immune response and are known to be genetically different[110]. One therefore can expect strong differential gene expression between the two. In fact, because of this expected difference, most analyses on this dataset have been performed using only one of the two types of ALL (e.g., [1],[111]).

Figure 5.7 shows the hierarchical clustering of the ALL data using both a Euclidean (panel a) as well as a Pearson distance (panel b). Remarkably, the clustering based on Euclidean distances does not cluster the two types of cells. We searched for causes for the unexpected clustering by coloring the labels by the various phenotypic variables available (data not shown), but without a result. If we cluster the same data using the same linkage function but with Pearson correlation as distance matrix, the picture changes drastically (Figure 5.7b). Here, the samples fall in two predominant clusters that correspond perfectly with the type of the ALL cell.

This example clearly indicates that the use of Pearson correlation distances generates clusters that seem biologically more relevant compared to clusters based on Euclidean distances.

5.4.2.2   Clustering algorithms

The term "clustering" applies to a wide variety of unsupervised methods for organizing multivariate data into groups with roughly similar patterns [87]. They can be roughly subdivided into two categories: hierarchical clustering methods[113] and partitioning methods[114]. Hierarchical algorithms find successive clusters using previously established clusters. Partitioning algorithms seek to minimize the heterogeneity of the clusters (within-cluster dis-

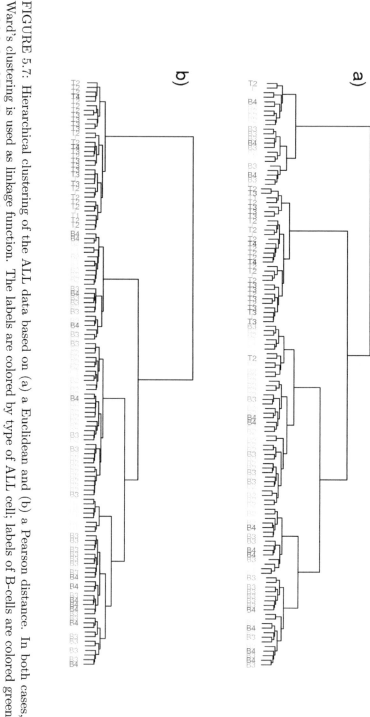

FIGURE 5.7: Hierarchical clustering of the ALL data based on (a) a Euclidean and (b) a Pearson distance. In both cases, Ward's clustering is used as linkage function. The labels are colored by type of ALL cell; labels of B-cells are colored green and labels of T-cells are colored red.

Re-sampling techniques are computer intensive statistical methods that use the samples in a (microarray) dataset multiple times. They have the advantage that they do not rely on idealized assumptions, like linear relations or normal distributions. Re-sampling techniques come in very different forms and address very different issues in microarray data analysis. This heterogeneous group can largely be subdivided in three very distinct types of statistics:

1. *Jackknife and bootstrap* to estimate the precision of calculated statistics like averages or variances. This is typically done by calculating the statistic of interest multiple times on subsets of available data (see StatsBox 5.23 for the difference between the two).

2. *Permutation tests* to build the null hypothesis distribution of a statistic, for example the distribution of the t-test statistic if there would not be a difference between the two groups. This situation of no difference can be simulated by re-sampling while randomly exchanging the labels of the samples (see Section 5.5.2.3 for more details).

3. *Cross-validation* to validate models by using random subsets. This is typically done by re-sampling whereby some samples are left out at each re-sampling iteration. These samples are then later used for validation or testing (see Section 5.6.2.3 for more details).

Re-sampling is typically done on the sample level and not on the gene level. The general aim of re-sampling methods is to construct many hypothetical replicates of an experiment based on the data of a single experiment. As replicated experiments would contain different sets of samples, it is logical that they are the samples that need to be re-sampled. Permutation tests of genes are particularly dangerous as they assume that the genes are independent, which is clearly a false assumption[112].

StatsBox 5.4: Re-sampling techniques

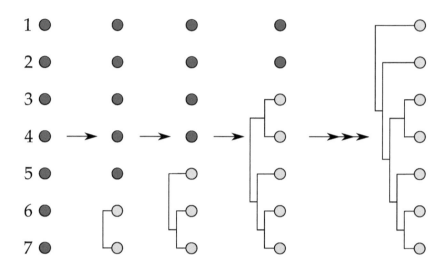

FIGURE 5.8: Schematic outline of a hierarchical clustering algorithm. This bottom-up approach iteratively groups in each step the two most similar clusters together. In the above sample, 6 and 7 are most similar. In the second step, 5 is most similar to 6 and 7. The following step identifies 3 and 4 to be most like 5, 6 and 7. This process is repeated until all clusters are grouped. The dendrogram visualizes the result of this grouping process.

tances) and/or to maximize their separation (between-cluster distance)[115]. The most commonly used partitioning algorithms are k-means[116], partitioning around medoids (PAM; a.k.a. k-medoids)[114] and self-organizing maps (SOM)[117],[118].

5.4.2.2.1 Hierarchical clustering Hierarchical clustering[113] generates a tree where the branches contain similar samples or genes. It is the first, and probably the most widely applied, method for clustering genes and samples in microarray data[119]. It starts by considering the n genes/samples as n clusters. At each iterative step, the two most similar clusters (i.e., with the shortest distance) are agglomerated into new cluster (see Figure 5.8). This is the bottom-up approach to form clusters, called the "agglomerative method." It is most commonly used, although there is also a top-down alternative, called the "divisive method," where the clusters are successively separated into smaller groups.

Hierarchical clustering results are mostly visualized by a dendrogram, a diagram with a tree structure, which is often used in combination with a heatmap (see Section 6.1.1). In such a dendrogram, genes with similar expression profiles across samples are located next to each another. A further effect of this sorting is the generation of groups (clusters) of genes which share a similar profile.

Pros and cons. After n-1 steps, the data have been structured into a tree structure. Tree structures are easily viewed and understood[87]. The methods do not require estimation of the number of clusters, but one can break down the tree into a desired number of clusters by cutting the tree at a certain height. The choice for this height remains subjective and hence arbitrary, but one can take multiple cut-offs thanks to the hierarchical nature of the tree.

There are, however, some intrinsic difficulties with hierarchical clustering. The decision to merge two clusters at a certain step is based on pair-wise distances of all clusters at that particular step, and not based on a "global" criterion[119]. Once a mistake is made, there's no mechanism or global evaluation to recover it in later steps[114]. When n is large, accumulation of mistakes is pronounced and the method lacks robustness.

5.4.2.2.2 K-means clustering This is a classical and widely applied clustering method[116]. It is basically an optimization algorithm that searches for an optimal allocation of the samples to a pre-specified number of clusters so that the total within-cluster variance is minimized. First, the numbers of clusters (k) is specified by the user, and the centroids (i.e., multidimensional center points) of these k clusters are initialized. Then, the samples are assigned to the nearest centroid's cluster and the centroids moved to a new location, in such a way that the sum of squared distances within the clusters (see equation below) is minimized (see Figure 5.9). These latter two steps are iterated until convergence, i.e., no movement anymore of the cluster centroids. K-medians also exists, which is highly similar to k-means but uses the more robust median instead of the average for computing the centroids.

$$\text{variance}_{\text{within clusters}} = \sum_{j=1}^{k} \sum_{x_i \in C_j} d_E(x_i, \overline{x}_j)$$

where there are k clusters $\overline{x}_j$($j = 1, 2, ..., k$), where $\overline{x}_j$ is the mean of all the $x_j \in C_j$ and where d_E is the Euclidean distance (see Section 5.4.2.1.1).

Pros and cons. Like most optimization algorithms on complex data, k-means may fall into a local minimum depending on the choice of the initial values. A common way to avoid such local minimum problem is to run the k-means algorithm multiple times with different initial cluster centers. Afterwards, the "best" cluster solution is selected, i.e., the solution with smallest within cluster sum of squares[119]. As an algorithm of global criterion, k-means usually produces good clustering results if k is correctly chosen. How-

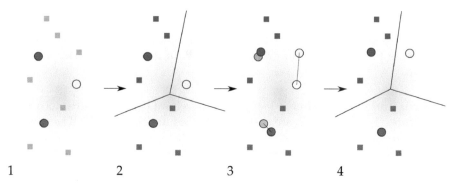

1   2   3   4

FIGURE 5.9: Schematic outline of the k means algorithm for the two-dimensional case. The first step randomly places the centroids of a user-defined number of clusters. The second step assigns each sample to a cluster based on the closest distance to a centroid. During the third step, the centroids are repositioned so that the sum of the squared distances within all clusters is minimized. Step 4 symbolizes the repetition of steps 2 and 3 until no further movement of the cluster centroids occurs.

ever, as with most clustering methods, k-means is quite unstable and highly affected by the presence of scattered genes in the complex microarray data[119].

The need to specify a priori the number of clusters k is probably the major drawback of k-means. An inappropriate choice of k may yield poor results, and the need for this specification contradicts somewhat with the method's philosophy of being unsupervised. How to choose the number of clusters k? There are some heuristic approaches among which re-sampling techniques to compare the quality of clustering results for different values of k[120].

5.4.2.2.3 Partitioning around medoids

Partitioning around medoids (PAM)[114], also known as k-medoids, is very similar to k-means. It makes use of the median centroids instead of the mean centroids, and is not restricted to the Euclidean distance but can be used with any distance measure. It makes use of an iterative procedure analogously to k-means in order to minimize the within-cluster variance, calculated as

$$\text{variance}_{\text{within clusters}} = \sum_{j=1}^{k} \sum_{x_i \in C_j} d_*(x_i, \tilde{x}_j)$$

where there are k clusters $\tilde{x}_j$ (j = 1, 2, ..., k), where $\tilde{x}_j$ is the median of all the $x_j \in C_j$ and where d_* can be any of the distance matrices specified in Section 5.4.2.1.1.

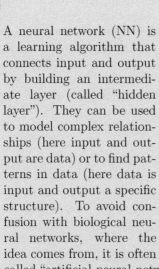 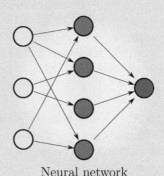

A neural network (NN) is a learning algorithm that connects input and output by building an intermediate layer (called "hidden layer"). They can be used to model complex relationships (here input and output are data) or to find patterns in data (here data is input and output a specific structure). To avoid confusion with biological neural networks, where the idea comes from, it is often called "artificial neural network (ANN)."

Neural network

In SOM for example, the interconnecting artificial neurons are the cluster centroids (i.e., the nodes of the grid), the input is the microarray dataset and the output is the grid space (see text for more details).

StatsBox 5.5: Neural nets

Pros and cons. Clustering results of PAM are usually similar to k-means, and it therefore shares the same advantages and disadvantages. However, PAM is more robust compared to k-means.

5.4.2.2.4 Self-organizing maps A self-organizing map (SOM)[117],[118] is a type of artificial neural network (see StatsBox 5.5) that produces a low-dimensional (typically two-dimensional), discretized representation of a microarray dataset, called a "map." The map seeks to preserve as many properties of the original microarray data as possible.

SOM first maps k cluster centers (called nodes) in a low-dimensional (usually two-dimensional) grid space from the high-dimensional space that the dataset is situated and then the cluster centers are adjusted iteratively. Each time, a point from the data is randomly chosen, and all cluster centers are moved towards this selected point. The closest cluster center will move the most, the more distant nodes will move less. The magnitude of movement decreases per iteration and stops after convergence (no movement between two iterative steps).

Pros and cons. Self-organizing maps are generally considered sub-optimal compared to k-means because the optimization is restricted on the two-dimensional node space[119]. On the other hand, if two dimensions are sufficient, self-organizing maps have the advantage that they preserve more characteristics of the original gene expression data compared to k-means. One however needs to be careful when interpreting well-conserved patterns of entire microarray data. Some of the intensity variation in microarrays is expected to be biologically irrelevant due to the noise of the methodology. Again, similar to k-means and PAM, SOM is very sensitive to the choice of the number of clusters[119].

5.4.2.2.5   Model-based clustering Model-based clustering methods will not force noisy genes to be part of a cluster (so-called scattered genes). It is indeed expected that many genes in an microarray study are biologically irrelevant (see Section 5.3). Forcing all these genes into cluster formation can introduce more false positives and distort the structure of identified clusters[119]. Model-based clustering[121],[122] is based on a mixture of several normal distributions and an additional Poisson distribution for the scattered genes. In other words, it defines clusters as sub-populations with certain (normal) distributions. Tight clustering[123] utilizes a repeated re-sampling approach to provide better robustness and directly searches for tight clusters.

Pan et al.[124] indicated that model-based clustering of t-statistics (or other statistics) can be a useful to exploit differential gene expression for microarray data.

Pros and cons. Model-based clustering provides a solution to the problem of obtaining spurious clusters due to noisy genes. It also allows to estimate the number of clusters and will generally result in more stable clusters[124]. Compared to the above mentioned clustering methods, it has a statistically more solid foundation and is consequently a more suitable framework for inference and interpretation. However, because of the high-dimensional and complex structure of the microarray data, this methodology still may result in spurious clusters due to errors in the selection of distributions and due to local minima in the optimization phase[119].

5.4.2.2.6   Biclustering All clustering methods described so far have two major limitations in common which are addressed in Biclustering methods [125]. First, they are applied to either the genes or the samples of the microarray data separately, but not simultaneously. Second, clustering methods derive a global model. Samples are clustered based on the activity of all the genes, and gene clusters are defined using all samples. It is, however, well known that genes may be co-regulated and co-expressed only under specific experimental conditions. While such local expression patterns may be missed using the clustering methods described above, biclustering is (theoretically) capable to discover them.

Biclustering is a family of quite different techniques (e.g., [126],[127],[128],[129],[130],[131],[132],[133]) that all allow a simultaneous clustering on the row and column dimensions of the gene expression matrix. These methods can identify groups of genes showing similar activity patterns under a specific subset of the studied samples. Biclustering is a field in development, with a plethora of techniques having quite distinct foundations. Despite some nice efforts like the thorough comparison study of Prelic et al.[130], it is therefore too early to claim which method outperforms the others. Given the subtlety, the complexity and the potential heterogeneity of the demanded patterns on the one hand and the big differences between methods on the other hand, it is even expected that some methods will be more advisable under certain conditions.

Pros and cons. The main advantage of biclustering is that it addresses an important and challenging question, as it can identify subsets of genes that are co-regulated under certain experimental conditions. There is however no such thing as a free lunch. Trying to discover such local patterns from a complex high-dimensional dataset in an unsupervised way involves practical difficulties.

One issue is which method to choose. Just like the choice of distance and linkage measures has profound effects on clustering outcomes, the obtained biclusters will depend on certain criteria intrinsic to the biclustering method used. While it is generally agreed upon that Pearson correlation (or Euclidean distance for relative values) are appropriate choices for hierarchical clustering, there is no consensus yet for biclustering. Again, given the complexity of the demanded patterns there probably is no one-size-fits-all solution.

Another issue is interpretability. It is difficult to interpret the many biclusters that are typically extracted from microarray data[125]. First, getting an overview of the different biclusters and what information they contain is not straightforward[134]. Second, and more importantly, understanding which biclusters are biologically most relevant (and why) can be a laborious task. The simplicity of biclustering, in the sense that it does not require assumptions or hypotheses, will often turn out to be deceptive. One doesn't need to reflect about well-defined hypotheses before the analysis, but will eventually have to do it at the end. And ironically, these raised hypotheses need to be verified, i.e., tested, in a follow-up experiment. Otherwise one walks into the trap of overfitting (see Section 5.2.2).

5.4.2.3 Quality check of clustering

Clustering methods can be quite unstable and will always produce partitions, even with random data. It is therefore wise to check the quality and the robustness of the obtained clusters. This can be done internally by re-sampling techniques and by silhouette graphs or externally by investigating independent test samples.

Several re-sampling techniques have been applied to validate the clustering robustness and to provide more stable clustering[120]. Silhouette functions are used to evaluate the strength of cluster membership by calculating the relative magnitudes of within- and between-cluster proximities. And a typical final quality check is typically done in the interpretation of the clustering; clusters without any biological meaning will not pass this check.

5.4.3   Multivariate projection methods

A multivariate projection method reduces the dimensionality of microarray data while retaining as much information as possible. This is done by transforming the data to a new coordinate system so that the largest source of variation of the data lies in the first coordinate (called the first principal component), the second largest variance in the second coordinate, and so on. All principal components explain additional variation compared to one another, meaning that they are completely uncorrelated.

Multivariate projection methods are especially useful for taming the collinear nature of microarray data, originated by the fact that many genes are co-regulated or operate in the same pathway. Correlated genes are combined into a single independent component. This lies at the heart of the methodology; it is by incorporating all correlated genes into a single component that the dimensionality is being reduced.

The obtained components can be used for further analysis, but are mostly used for data visualization. Typically, the first two components are shown in a two-dimensional scatterplot (see Section 5.4.3.2). As these components represent the two largest independent sources of variation present in the data, such a figure actually reveals the most informative part of the internal structure of the microarray data. Of course, *much* information is not necessarily equal to *relevant* information. If the comparison of interest, say the effect of an experimental treatment, is not dominating the variation in the data, it might not be observable in the first components. On the other hand, in such a situation, it might be useful to know which sources of variation are dominating the effects of treatment.

5.4.3.1   Types of multivariate projection methods

In the algorithm of multivariate projection methods, different building blocks can be distinguished (see StatsBox 5.6). There are three methods that are widely used for multivariate projection:

1. Principal component analysis (PCA[135])

2. Correspondence factor analysis (CFA[136],[137])

Main steps in multivariate projection methods [139].

Transformation. A recalculation of the data so that the distribution of its values is transformed, mostly to approach a normal distribution.

Closure. Dividing each value of a set of values by the total sum, so that the sum after closure equals 1.

Centering. Subtracting the mean of a set of values from each value, so that the mean after centering equals 0.

Scaling. Dividing each value of a set of values by the root mean square, a measure of variation, so that the standard deviation after scaling equals 1.

Factorization. The decomposition of a dataset into a product of other objects, called factors, which when multiplied together give the original. Singular value decomposition (SVD) is such a factorization model, which is typically being used for multivariate projection methods[140].

Projection. Plotting the first few factors on a graph, called biplot, that displays both genes and samples. There are several ways to scale the factors of the genes and samples before plotting them on the same graph.

StatsBox 5.6: Steps of multivariate projection methods

3. Spectral map (SPM[138])

They have several differences in more than one of the building blocks shown in StatsBox 5.6. For a comprehensive overview of all steps, and a more detailed explanation of the mathematics behind, we recommend the paper of Wouters et al.[139]. Here, we will discuss only the building blocks that are responsible for the main differences between these three methods, together with their main advantages and disadvantages.

Pros and cons. The first issue in multivariate projection methods is the transformation of the data. Log-transformation is optional in PCA and absent in CFA, but is default in SPM. As log-transformation is beneficial for microarray data for multiple reasons previously discussed (see Section 4.1.2), we prefer to log2 transform the data.

The second issue is whether the gene expression profiles should be examined in absolute or in relative terms. PCA typically shows the absolute gene

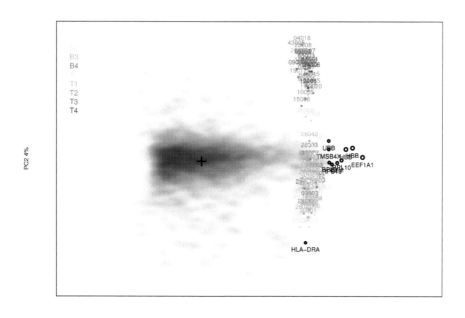

PC1 73%

FIGURE 5.10: A biplot of the first two components of a PCA shows a discrimination of the two types of ALL (B-cell and T-cell) in the y-axis. Squares represent the samples, and circles represent the genes.

expression levels by only centering and scaling the samples (the columns), but not the genes (rows). CFA and SPM both look at the data in a relative way, but based on a different algorithm; CFA removes the size component from the data by closure of the samples and genes, and SPM by centering and scaling both samples and genes (see StatsBox 5.6). Looking at relative gene profiles makes more sense in microarray data, as we are mainly interested in contrasts between samples and not in the absolute degree of intensity. A typical PCA of the ALL data, for example, will discriminate the two types of ALL (B-cell and T-cell) in the second component (Figure 5.10). The first component is dominated by biologically irrelevant variation due to variation in absolute intensities. A SPM of the same data, in contrast, directly discriminates the two groups in the first component (Figure 5.11).

A third major issue is the scaling of the factors of the genes and samples for projection on the biplot (see StatsBox 5.6). CFA typically scales genes and samples symmetrically, SPM asymmetrically, and PCA can do either way. Because of the large difference in row and column dimension in microarray

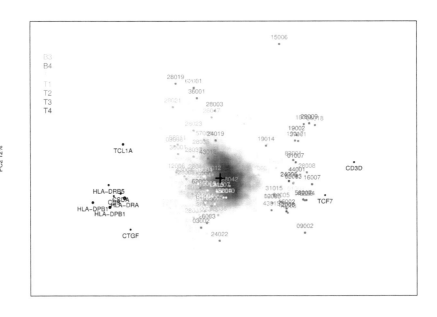

FIGURE 5.11: A biplot of the first two components of a SPM shows a clear discrimination of the two types of ALL (B-cell and T-cell). Squares represent the samples, and circles represent the genes.

data, an asymmetric factor-scaling is recommended to construct the biplot. Such an operation pulls the genes away from the center of the biplot while leaving the samples at their original places.

Because SPM provides appropriate solutions to each of these three above-mentioned issues, its use is recommended for projecting microarray data. A further useful property of SPM is that the size of the symbols in the biplot is made proportional to the mean intensities of genes and samples. This procedure elegantly reintroduces the size component that was previously removed by the centering. Hence, the size of a gene symbol will therefore reflect its average expression level.

Spectral map also has many advantages compared to the other clustering methods described in Section 5.4.2. In contrast to hierarchical clustering, it uses a "global" and well-defined criterion (i.e., explaining maximal variance in the data). In contrast to k-means and SOM, it does not require to specify in advance the number of expected clusters. The biplot neatly visualizes both samples and genes on the same plot, so that it is straightforward to deduce which genes are responsible for which clusters. As there are as many components as there are genes, SPM allows the detection of multiple types of groupings depending on which components were used.

SPM also preserves the quantitative nature of the gene expression levels. The SPM of the MAQC study (see Figure 4.12) is a nice example. The MAQC consortium selected two distinct reference RNA samples (A and B), and mixtures of these two samples at defined ratios of 3:1 (C) and 1:3 (D). The x-axis of the spectral map indeed shows a clear clustering of the four sample types A, B, C and D (see Section 4.2.4 for a discussion on why the samples of site 6 are discriminated from the others). Interestingly, the samples of type C are positioned 1/4 to type A and 3/4 to type B. The samples of type D, conversely, share only 1/4 of the variation with type A but 3/4 with type B. The position of the samples on the SPM therefore perfectly reflects the quantitative proportions of the 3:1 and 1:3 ratios of sample types C and D.

5.4.3.2   Biplot

The first components of a multivariate projection method are typically displayed in a biplot that combines the genes and samples in the same plot. By coloring the samples one-by-one by the phenotypic variables at hand, one can rapidly reveal the main factors underlying the structure in the data.

Squares represent the samples, and circles represent the genes. The size of the circle corresponds to the average absolute intensity of that gene. Distances between squares are a measure for similarity between samples, and analogously for genes. A positive association of a gene with a given sample (i.e., an up-regulation of that gene in that particular sample) results in the positioning of the gene and sample on a common line through the centroid (depicted by a cross). In Figure 5.11, for example, the genes on the right and the left of the graph are the genes with the highest contribution for the separation between T- and B-cells. More specifically, genes in the same direction as the T-cell samples are over-expressed in T-cells compared to B-cells. Indeed, the two genes (TCF7 and CD3) are well known to be specifically expressed by T-cells[141],[142]. The genes on the same side as the B-cells (like HLA-DMB, HLA-DRB5, etc.) are all histocompatibility class II molecules. Interestingly, it is well known that this type of molecule is normally expressed on only certain types of cells, including B lymphocytes, but not on T-cells (in contrast to class I which are expressed in both cell types).

Regularly, some straightforward clusters appear that explain the most variation in the data and are consequently plotted in the first axes of the spectral map. If there are male and female samples in the study, the spectral map often clusters the two genders because of the large variation between them in the expression levels of several Y-linked genes (higher in males) and the expression of the inactivating X gene (Xist, higher in females). A logical thing to do is to ignore the first principal component and plot the subsequent ones.

Another valuable approach that we would recommend is to apply a spectral map analysis on each of the clusters, like in the example a spectral map for the males and females separately.

5.5 Detecting differential expression

A common objective in microarray studies is to identify the genes that are consistently differentially expressed under certain conditions. To this end, one estimates the difference in expression levels between the conditions, and one tests whether the observed differences are statistically significant. This is typically done *gene-by-gene separately*. Although such an approach ignores the omnipresent dependencies between genes, it is very helpful in extracting relevant information from the high-dimensional and complex microarray data. Afterwards, all the separate tests can be aggregated again by ranking the genes based on statistical significance and/or absolute difference in expression values.

Detecting differential expression between conditions may come in very different ways. First, it depends on the *choice of statistic*, which in turn depends on assumptions on how expression levels are believed to be distributed across the samples. Second, it depends on the *design of the study*. There may be two experimental conditions or many, the conditions may be independent or related to each other in some way (as in a time series), or there may be many different combinations of experimental variables. Replicates, if present at all, might be samples from different animals or repeated hybridizations of the same samples. Reflecting this variety, many different methods are commonly used for identifying significant changes[87]. We will show here that most of these approaches can be addressed by derivations of the same basic model.

An important issue in detecting differential expression is the problem of multiple testing. Because of the gene-by-gene approach, a large number of tests are carried out (as many as there are genes), which generally results in a large number of false positive findings. Multiple testing correction and the choice of significance levels are crucial to determine a cut-off to claim which genes are significant and which are not, but they do not change the ranking of the genes (see StatsBox 5.7).

5.5.1 A simple solution for a complex question

A microarray enables to simultaneously measure the expression levels of tens of thousands of genes. All these expression levels are measured in the same cells at the same moment. As genes work in pathways and in combination, their expression levels are clearly not independent. Expression data

Steps in determining differential expression[143].

1. Select a statistic. This statistic is used to rank the genes by order of evidence for differential expression. The choice of statistic can have a big impact on the obtained results, especially in small sample-sized studies.

2. Choose a cut-off value. This arbitrary value is used to consider all genes with a higher statistic (or lower p-value, see Section 5.5.2.3) as being significant. This cut-off will primarily depend on the preferred confidence level (significance level) and the number of tests performed (impact of multiple testing correction).

The first step is the most important, and luckily also the easiest. The second step involves an **arbitrary** choice by which one cuts the ranked genes into a significant and a non-significant group. The danger of this decision step is that some top-ranked non-significant genes may be more relevant than some significant genes.

StatsBox 5.7: Steps in determining differential expression

therefore typically contain genes that are correlated or anti-correlated. Furthermore, as feedback loops are omnipresent and as certain genes may inhibit or catalyze the expression of other genes, some gene pairs may turn out to be very interesting when analyzed in combination.

Tests for differential expression are univariate, i.e., they treat each gene separately. As hypothesis testing ignores the multivariate nature of gene expression, it comes across as an incomplete approach in theory. It ignores the existence of correlations, and it will fail to discover interesting gene combinations. A pertinent question is consequently why this overly simplistic approach is the most commonly used analysis method in gene expression studies.

The incomplete approach of testing single genes independently for differential expression works remarkably well in practice. The main reason is because of the curse of high dimensionality of microarray data (see Section 5.2). Because there are typically so many groups and subgroups of correlated and interacting genes present in a single microarray, a multivariate approach that tries to incorporate most or all covariances becomes highly complex. And such a complex model has two drawbacks: it is difficult or impossible to interpret, and it is likely to overfit thereby impeding its generalizability (see Section

5.6.2.1). In a complex and high-dimensional context, a simple solution will often guide the researcher to the clearest and most informative patterns of the data.

Note that the independently obtained results (p-values and fold changes) are aggregated for all genes at the end in typical lists and plots. Imagine a situation where 50 genes that are highly correlated are all strongly differentially expressed. These 50 genes will all be found at the top in a table containing all genes ranked from most to least significant p-value. In this way, one investigates the entirety of genes, although in an indirect and incomplete way.

We personally experience that both univariate or multivariate views on the data have their merits and are actually quite complementary. A thorough analysis approach should therefore include both views.

5.5.2 Test statistic

There are many test statistics to identify the genes that are differentially expressed between conditions. The choice of test statistic can greatly affect the set of genes that are identified, particularly in small sample-sized studies (see Figure 5.18). Despite the wealth of available methods, biologists often show a fondness for two of the earliest approaches, fold change and the t-statistic, presumably because of their simplicity and interpretability[144]. Many authors[144],[145],[146] have however shown that the new generations of t-statistics, which are modifications from the ordinary t-statistic, improve the quality of the obtained top gene lists.

5.5.2.1 Fold change

A fold change (FC), or log ratio, is the difference per gene between the averages of the two groups in log2 scale. For instance, if the average expression levels of the treatment were subtracted from the control condition, then genes with fold changes of -1 and 3 were respectively regulated 2-fold down and 8-fold up by treatment.

In early studies, genes were often ranked with respect to fold change. Genes showing fold change above 2 (or another arbitrary cut-off) were regarded as potentially regulated and were selected for further investigation. The obvious drawback with such an approach is that genes with high fold change may also be highly variable and thus with low significance of the regulation[147].

Some recent papers on the MAQC study have indicated that gene lists based on fold change ranking are more reproducible than based on ordinary and modified t-statistics[148],[34]. A reproducible gene list is however not equal to an accurate or a relevant gene list[144]. Whether fold change or a (modified) t-statistic results in a more appropriate and accurate gene list depends on the research question. An interest in absolute changes in gene expression suggests the use of fold change, while a t-test is more appropriate if interested in changes in gene expression relative to the underlying noise in the gene. there-

Log of the ratio is equal to difference between the logs.

$$log_2(\frac{A}{B}) = log_2(A) - log_2(B)$$

This is why we search for differences in a log2 scale to discover interesting patterns in the ratios of gene expression between groups. The fortunate side effect is that analyses are much easier to perform in an additive scale than in a multiplicative scale. Testing for a difference in average gene expression between group A and B in a log2 scale, for example with a t-test, answers whether the ratio of gene expression levels of A over B is significantly large or small. All linear models (see Section 5.5.2.5) are additive models and can be performed after log2 transformation of the data.

StatsBox 5.8: log of the ratio = difference between the logs

fore be based on biological, rather than statistical, considerations[144]. In the case of the MAQC study, Shi et al.[148],[34] should have pointed out that "concordance between gene lists" is not necessarily applying to "all the truly differentially expressed genes" but rather to the genes whose large fold change could be detected no matter what measuring method or data pre-processing algorithm is used[149]. The MAQC compared very disparate mRNA populations (total human RNA vs. human brain RNA) while using technical replicates. As most microarray experiments make use of biological replicates and compare much more similar mRNA populations, it is more appropriate to check whether the subtle differences exceed biological variation than to check for large fold changes[149].

Two remarks should be made on the dilemma to choose between fold change and significance testing. First, a volcano plot (see Section 6.1.5.1) combines the fold change information together with p-values. As a volcano plot visualizes the rankings of both the t-statistic and the fold change, it is an invaluable tool for a better understanding of the data. Second, the second generation of moderated t-statistics (see Section 5.5.2.2) are somewhere in between a fold change and a t-statistic.

5.5.2.2 Types of t-tests

Why is a fold change insufficient? Figure 5.12 shows three possible outcomes in a study comparing two groups. The fold changes are exactly the same in

Null hypothesis. Null hypotheses are typically statements of no difference or effect. The null hypothesis is presumed true until statistical evidence, in the form of a hypothesis test, indicates otherwise[a].

P-value A p-value measures the strength of evidence against the null hypothesis, as it is the probability of observing this result, or more extreme, if the null hypothesis is true. A p-value close to zero indicates that the null hypothesis is false.

For example, a p-value of 0.001 indicates that, if there is actually no effect or difference, the observed test statistic would only occur once every 1,000 times.

The smaller the p-value, the more convincing the evidence is against the null hypothesis. What is "small enough" needs to be defined by the researcher. If the p-value is smaller than this pre-specified significance level (often 0.05), the null hypothesis is rejected; the effect or difference is then called "significant" (at 0.05).

---

[a]Conceptually, hypothesis testing is like a deliberation of a jury in a court. The null hypothesis is assumed true until the data argue clearly to the contrary.

StatsBox 5.9: Null hypothesis and p-value

all three situations. The only thing that differs is the variability or "spread" in observed expression levels around the group means. Despite having exactly the same fold changes, the two groups are not equally "different" in the three situations. This illustrates that a scientifically solid search for differential expression should not only imply subtracting one mean from the other, but should also take the variability around the means into account. A small difference between means will be hard to detect if there is lots of variability or noise. A large difference between means will be easily detectable if variability is low. This way of looking at differences between groups is directly related to the signal-to-noise metaphor; differences are more apparent when the signal is high and the noise is low[147].

5.5.2.2.1 T-test

What is the standard statistical approach to test for a difference between groups A and B? Estimate the difference between the two groups $(\bar{x}_A - \bar{x}_B)$, which is the same as the fold change $\bar{M}$, estimate the variation in the dataset (variance s), and check whether the signal-to-noise

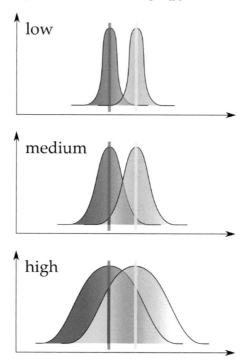

FIGURE 5.12: Three possible outcomes in a study comparing two groups. Despite having exactly the same fold changes, the two groups are not equally "different" in the three situations, hence leading to different significance levels.

ratio is exceeding a certain threshold level (see Figure 5.13). This threshold level depends primarily on the sample size (see Section 5.5.2.3).

The formula underlying the t-statistic is shown below. It depends on three parameters:

1. **The fold change ($\bar{M}$).** A larger $\bar{M}$ results in a larger t, indicating that genes with larger differences in expression are more significant (if having the same variance and sample size).

2. **The variance (s).** A larger s results in a smaller t, indicating that genes with an identical fold change but with smaller within-group variation are more significant.

3. **The sample size (n).** A larger n results in a larger t, indicating that larger sample-sized studies contain more significant genes despite having the same number of genes, the same fold changes and the same gene-specific variances. This is logical, as the estimates of signal and noise become more accurate and representative when based on more samples. This explains the increased power of studies with larger sample sizes;

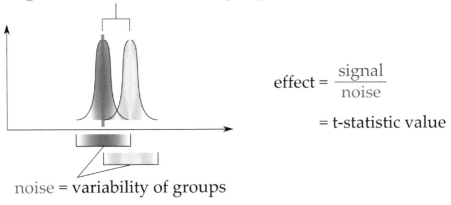

signal = difference between group means

$$\text{effect} = \frac{\text{signal}}{\text{noise}}$$

$$= \text{t-statistic value}$$

noise = variability of groups

FIGURE 5.13: Schematic view of how data lead to a t-statistic value. The difference between the two groups and the variability within the groups is estimated, and their ratio is the t-statistic. Hence, a t-test provides a measure for the signal-to-noise ratio.

the same difference can be called significant only because more samples have been used.

$$Student's\ t = \frac{\bar{x}_A - \bar{x}_B}{\sqrt{s^2/n}} = \frac{\bar{M}}{s/\sqrt{n}} = \bar{M} \times \sqrt{n} \times \frac{1}{s}$$

An ordinary t-test assumes the two groups to have equal variances. A safer approach in microarray studies is to allow different estimates for the two variances. This can be done with the Welch's t-test, an adaptation of Student's t-test for when the two groups possibly have unequal variances. In this approach, one does not calculate one overall variance s, but separate variances for groups A and B (s_A and s_B).

$$Welch's\ t = \frac{\bar{x}_A - \bar{x}_B}{\sqrt{\frac{s^2_A}{n_A}} + \sqrt{\frac{s^2_B}{n_B}}}$$

Since the number of replicates in many studies is small, variance estimators computed solely within genes are not reliable in that very small values can occur just by chance[144]. As a consequence, the ordinary t-test suffers from low power and is not the best option for detecting differentially expressed genes.

5.5.2.2.2 Penalized *t*-test The penalized t, often called d-value[150], is calculated like an ordinary t-test but with the estimate of the variation s augmented with a positive constant, the fudge factor s_0 (see equation below).

Variance, standard deviation and standard error.

The **variance** (s^2) and the **standard deviation** (s) are parameters or statistics that describe the dispersion of the values in a population. It is calculated by the sum of squares (SS) of the deviations from the mean:

$$s^2 = \frac{1}{n}SS = \frac{1}{n}\sum_{i=1}^{n}(y_i - \bar{y})^2$$

where n is the sample size, $\bar{y}$ is the mean, and y_i is the observation for sample i.

The **standard error** (SE) balances the dispersion of a population with the error associated with the sampling process. It is the estimate of the expected error in the sample estimate of a population mean. In other words, it measures how precisely the estimate of the population mean (or another parameter) actually is. The standard error of the mean (SEM) is derived by dividing the standard deviation by the square root of the sample size.

$$SEM = \frac{s}{\sqrt{n}}$$

Larger sample sizes result in smaller standard errors, reflecting the fact that the estimate for the mean becomes more precise.

StatsBox 5.10: Variance, standard deviation and standard error

The primary goal of the term s_0 is to penalize genes with variances close to zero. These genes are likely to be called significant even when the difference is small. The small s of these genes makes the t-statistic large, irrespective of the magnitude of the fold change $\bar{M}$. Adding the fudge factor has a big impact on genes with small variances, but the effect will diminish with increasing variance as the fudge factor becomes proportionally irrelevant (see Figure 5.14).

$$Penalized\ t = d\ value = \frac{\bar{M}}{s_0 + s/\sqrt{n}}$$

Tusher et al.[150] estimate the fudge factor s_0 by minimizing a coefficient of variation of the absolute t-statistic, while Efron et al.[151] estimates it by 90th percentile of the distribution of sample standard deviations of all genes.

The penalized t as implemented by SAM (significance analysis of microarrays)[150] makes use of permutations to simulate for every gene a situation in which there is no difference between the two groups. First, the samples are randomly shuffled between groups A and B for a number of times (about 1,000 times) and afterwards the d-value is calculated in each of these permutated datasets (see StatsBox 5.4). The average of all these d-values is then used as an estimate of the expected d-value of that gene if it would not have been differentially expressed. The more the observed d-value of that gene deviates from the expected, the more likely it is that it is truly different between the two groups. For every gene, we end up with an observed and an expected d-value. Typically, the observed d-values are plotted vs. the expected d-values while ranking the genes on increasing expected d-value (see Figure 5.15). Next, an arbitrary cut-off (typically called delta) needs to be chosen. This choice is not that straightforward as it reflects the compromise one needs to make between the number of significant genes and the number of false positive results (see FDR in Section 5.5.3.2.2). A gene that deviates more than one delta from the diagonal, and all genes that have more extreme d-values than this gene are called significant (see Figure 5.15).

There are some other penalized t-tests besides SAM. Broberg[152] for instance proposed to determine the offset delta by minimizing a combination of estimated false positive and false negative rates. It is based on a nice rationale, but is computationally quite intensive and limited to comparisons between two groups.

5.5.2.2.3 Moderated t-test The moderated t-statistic for each gene is like an ordinary t-test, except that the standard error is moderated across genes, i.e. shrunk toward common value using an empirical Bayes model (see StatsBox 5.11).

This is done by replacing the gene-specific variance s from the ordinary t-statistic by the posterior variance $\tilde{s}$, which is a mixture of s and a global variance estimate s_0[153]. The relative contributions of s and s_0, respectively d

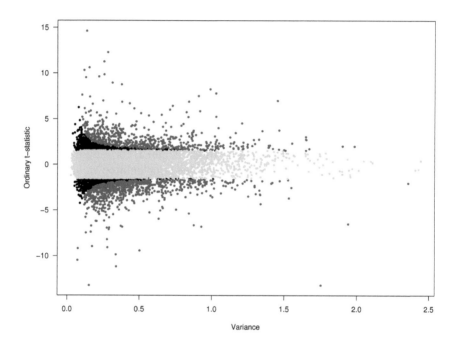

FIGURE 5.14: The penalized t-test: effect of variance on significance. A plot of the ordinary t-statistic vs. the variance of each gene. The non-significant genes are colored grey, the genes only significant with the ordinary but not with the penalized t-test are colored black and the genes significant with both the ordinary and the penalized t-test are colored red. Genes with an identical t-statistic but with smaller variances (on the left side of the graph) are less likely to be called significant with a penalized t. This graph is based on the comparison of 3 patients with the BCR/ABL fusion gene to 3 patients from the NEG group from the ALL dataset.

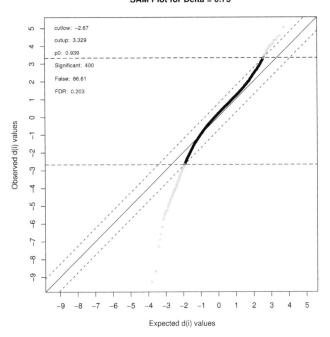

FIGURE 5.15: Plot of the observed d-values vs. the ordered expected d-values. Each gene is represented by a dot, and the differentially expressed genes are colored in green. This graph is based on the comparison of 30 patients with the BCR/ABL fusion gene to 30 patients from the NEG group from the ALL dataset. Compared to NEG, there are 9 genes being significantly upregulated (green dots above) and 131 genes being significantly downregulated in BCR/ABL (green dots below) at an FDR of 15%.

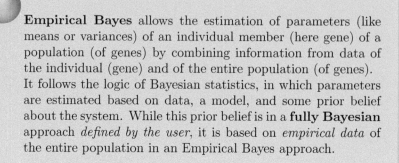

Empirical Bayes allows the estimation of parameters (like means or variances) of an individual member (here gene) of a population (of genes) by combining information from data of the individual (gene) and of the entire population (of genes). It follows the logic of Bayesian statistics, in which parameters are estimated based on data, a model, and some prior belief about the system. While this prior belief is in a **fully Bayesian** approach *defined by the user*, it is based on *empirical data* of the entire population in an Empirical Bayes approach.

StatsBox 5.11: Empirical Bayes

and d_0, are determined by the accuracy of the estimation of s. The moderated t reduces to the ordinary t if d_0 is zero and approaches a fold change if d_0 is very large[153].

$$Moderated\ t = \frac{\bar{M}}{\tilde{s}/\sqrt{n}} \ with \ \tilde{s} = \frac{s^2 d + s_0{}^2 d_0}{d + d_0}$$

The empirical Bayes shrinkage of the gene-specific variances toward a common pooled estimate has three main advantages compared to the ordinary t-statistic. First, it increases the degrees of freedom (see Section 5.5.2.3) for individual genes. The added degrees of freedom compared to the ordinary t-statistic reflect the extra information that is borrowed across genes for inference about each individual gene[153]. Second, it results in far more stable inference when the number of arrays is small. Third, like with SAM, the addition of a positive value to the gene-specific variance introduces a penalization of the moderated t-statistic for genes with small variances.

Some of the moderated t-statistics that make use of (empirical) Bayesian estimates are the approaches[154], Baldi and Long[155] and Smyth[153]. Baldi and Long[155] were the first to propose the idea of using a t-statistic with a Bayesian adjusted denominator in a program called "cyberT." Loennstedt and Speed[154] proposed the B-statistic, a log posterior odds ratio that estimates the likelihood whether a gene is differentially expressed or not. Smyth[153] expanded on their methodology and is currently the most widely applied implementation of the moderated t-statistic through the use of the R package Limma[156].

The moderated t-statistic as implemented in the R package Limma is particularly attractive because it extends naturally to more complex experiments than just two-group comparisons. For instance, it can test for more than two groups through the use of moderated F-statistics, and can test for interactions between the levels of two (or more) variables.

One caveat of Limma is that it does not use a local estimate of the standard deviation, but rather a single pooled estimate. If many genes have unrepresentative variances for the gene at hand, it might not provide such a relevant estimate. For that reason, one can for instance presume that the moderated t may work poorly on MAS 5.0 data without variance stabilization normalization [147].

A nuisance for many Limma users is that the coding is quite cumbersome as it is not trivial to specify the design matrices and contrasts. When struggling with this package, it is best to contact a colleague statistician.

Other tests There are many more tests to detect differentially expressed genes. To get a more comprehensive overview we recommend some comparison studies and reviews[145],[146],[147],[157],[158]. The concept of most methods can however be brought back to one of the three types described above (i.e., the ordinary, penalized or moderated t-statistic).

5.5.2.3   From t-statistics to p-values

The t-distribution is a probability distribution. As seen in StatsBox 5.9, the p-value is the probability of observing such a t-statistic by chance under the null hypothesis. The p-value of a t-test refers to the total area under a particular t-distribution from the absolute value of t to infinity (see Figure 5.16). In other words, we first calculate the difference in gene expression compared to within-group variation (the t-statistic). Next, we check how likely it is to observe such a t if the gene would not be differentially expressed. This is done by calculating the tail area under the t-distribution, which is the p-value (see Figure 5.16). The larger the absolute value of t, the smaller the p. Finally, a decision is made by comparing the observed p-value vs. an arbitrary significance level that reflects how confident one would be about his or her decision. A p-value below such level, say 0.05, indicates that the difference is real or significant in 95% of the cases.

The Student's t-distribution is not a unique distribution, but rather a family of distributions whose shape is determined by the degrees of freedom. The degrees of freedom is a measure of how much information an estimate or statistic has. A general rule is that the degrees of freedom increase when more samples have been measured, and decrease when more parameters have to be estimated. As the degrees of freedom increase, the shape of the t-distribution becomes steeper and approaches the normal distribution. This results in smaller areas from the absolute t value of t to infinity (see Figure 5.16), resulting in smaller p-values. Hence, the larger the sample size, the

Significance level and power.

The *significance level* is the cut-off probability to wrongfully reject the null hypothesis (false positive). In a microarray context, this is calling a gene significantly differentially expressed while it is actually not. The threshold for the p-value determines the probability of false positives, and defines when a gene is called "significant." In general, this threshold is set to 0.05. The *power* of a test is how soon a truly differentially expressed gene is indeed called "significant". In other words, the power of a test is the probability to correctly rejecting a false null hypothesis, which is equal to to $1 - proportion_{false\,negatives}$.

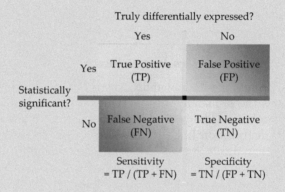

Significance level and power go back to the two possible errors made in a statistical decision process, the type I error or false positives (α) and the type II error or false negatives (β). A false positive result is in principle worse than a false negative result. The former gives false hope, while the latter indicates that the current study did not provide enough evidence to claim a true difference. The probability to make a false positive conclusion therefore needs to be specified by the researcher (the significance level). The probability of having true positive results (the power) depends on the sample size of the study, the degree of differential expression (the fold change), the noise in the data and the statistical method used.

Also sensitivity and specificity (see StatsBox 5.19) can be linked to type I and II errors.

StatsBox 5.12: Significance level and power

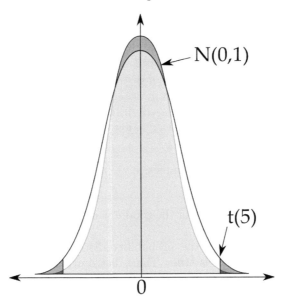

FIGURE 5.16: Two t-distributions with different degrees of freedom. A certain observed t-statistic is indicated together with its corresponding p-value, i.e., the total area under a particular t-distribution from the absolute t-value to infinity. The same t-statistic will result in a smaller p-value if there are more degrees of freedom.

more precise a statistic can be estimated so that the same statistic will result in a smaller p-value[1].

As mentioned in Section 5.5.2.2.3, the moderated t is shown to follow a t distribution with more degrees of freedom than the ordinary t-statistic. The added degrees of freedom compared to the ordinary t-statistic reflect the extra information which is borrowed from the ensemble of genes for inference about each individual gene.

A few more details are given on the t-statistic for the interested reader. A classical t-distribution is always symmetric, so that the p-value associated with t will always be the same as the p-value associated with t. The significance of differential expression will then be the same, but the direction opposite (up-regulation vs. down-regulation). There are also other frequently used distributions for other types of statistics, like the F-distribution for the ratio

---

[1]The t-test was derived in 1908 by William Gosset, who worked at a Guinness brewery in Dublin, to monitor the quality of beer brews. Gosset was not allowed to publish under his own name, as his employer regarded the use of statistics as a trade secret. Therefore, the paper was written under the pseudonym Student.

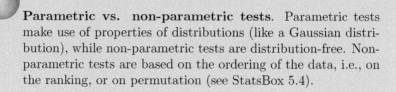

Parametric vs. non-parametric tests. Parametric tests make use of properties of distributions (like a Gaussian distribution), while non-parametric tests are distribution-free. Non-parametric tests are based on the ordering of the data, i.e., on the ranking, or on permutation (see StatsBox 5.4).

StatsBox 5.13: Parametric vs. non-parametric tests

of variances or the chi-square for the goodness-of-fit of an observed distribution to a theoretical one.

The null distribution is the probability distribution of the test statistic when the null hypothesis is true. The t-distribution under the null hypothesis, as mentioned previously, is therefore the same as the null distribution of the t-statistic. The t-test is a parametric statistic because it is assumed to fit a parametrized distribution, namely the Student's t-distribution. The non-parametric alternative is to derive the null distribution by permuting the samples and calculating the t-statistic for each permutation. As the permutation exchanges the labels across the groups, it simulates a situation where there are no differences between the groups, which is exactly the null hypothesis.

Another popular non-parametric alternative is the Wilcoxon rank sum test, also called the Mann-Whitney U test, who makes use of the ranks of the observations instead of the actual measurements.

5.5.2.4   Comparison of methods

Let's use the ALL data as a motivating example to compare the p-values of an ordinary t-test, a penalized t-test (SAM[150]) and a moderated t-test (Limma[153]).

First, we test for differential expression between the NEG and the BCR/ABL groups based on 30 samples in each group. Figure 5.17 shows the pairwise correlations between p-values obtained with a permutation t-test, an ordinary t-test, SAM and Limma, together with their relationship with the fold change. When comparing the permutated t-test and SAM (which is also permutation-based) with the ordinary t-test and Limma, it is clear that p-values generated using permutations tend to become more granular, particularly the very small p-values (orange ellipses in Figure 5.17). Except for the permutation t-test, the methods are highly correlated p-values (Pearson correlation of 1) and almost identical volcano plots (see Section 6.1.5.1 for how to interpret volcano plots). The main conclusion is that, if the dataset contains many samples, the choice of statistic does not matter much as all provide similar results.

Next, we do the same comparison but based on fewer samples, i.e., only 3 samples in each group. The results are shown in Figure 5.18. While SAM and Limma still have highly correlated p-values (Pearson correlation of 0.99), the ordinary t-test deviates from SAM (0.94) and especially from Limma (0.89). The genes with small variances, highlighted in red, are the cause of this deviations as the more variable genes, colored in blue, still follow the diagonal (see Figure 5.18).

The volcano plot of the ordinary t-test (top-right in Figure 5.18) neatly shows that quite a number of genes are significant despite a small fold change because of having a small variance (i.e., the red-colored genes in the orange ellipse). As can be seen from the volcano plot at the middle right, SAM does not generate such false positive genes because the variances were penalized. The volcano plot of Limma (bottom right in Figure 5.18) has even a larger gap in the volcano plot, indicating that small variance genes are even more penalized compared to SAM. The main conclusion is that, if the dataset contains few samples, the choice of statistic has a big impact on the results.

This comparison of methods based on the ALL data can be used to make some important remarks.

Sample size affects concordance between methods. There is no real difference between most methods when applied on data with many samples (say, more than 10-20 samples per group). The difference between methods increases however with decreasing sample sizes. In other words, one needs to carefully consider which test statistic to use when confronted with a study with few replicates (say, less than 10 samples per group). The choice of test statistic will affect the ranked gene lists and consequently the conclusions of the study.

Estimating gene variance is difficult with few samples. The increasing dissimilarity between methods with decreasing sample size is mainly due to difficulties with variance estimation. If the sample size is small, the estimates of variance are very unstable.

To appreciate the logic behind this important concept, imagine that one estimates the mean based on one sample. For example, a demographer who estimates the average body length of a male New Yorker based on only one individual. This estimate will be very unstable, as it will change almost every time another New Yorker is picked out. Using 100 males that are randomly selected in New York, in contrast, would make the estimate not only more representative but also more stable.

Estimating variance based on few samples is much more unstable compared to averages. Estimating how variable male New Yorkers are in body length requires more samples compared to estimating just their average body length.

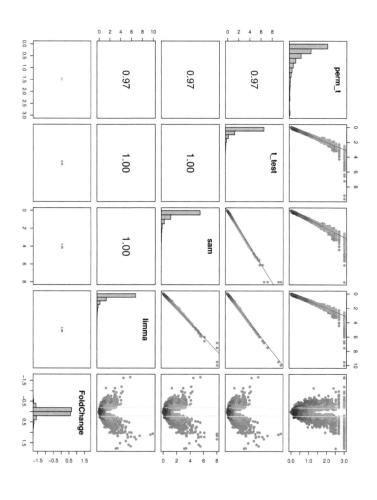

FIGURE 5.17: Comparison between methods to test for differential expression using a large sample-sized dataset (30 vs. 30). Pairwise correlations of the minus log10 transformed p-values, together with their respective Pearson correlation coefficient, are shown on the left. The red line indicates the equality line. Volcano plots are shown on the right. See text for more interpretation.

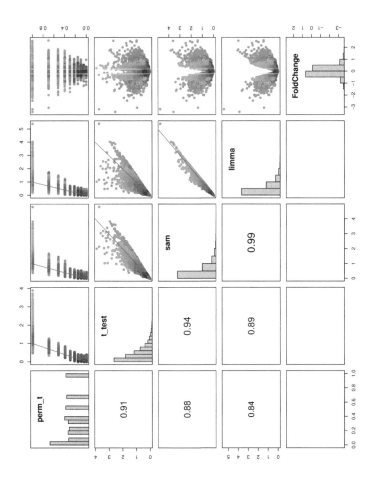

FIGURE 5.18: Comparison between methods to test for differential expression using a small sample-sized dataset (3 *vs.* 3). Pairwise correlations of the minus log10 transformed p-values, together with their respective Pearson correlation coefficient, are shown on the left. The red line indicates the equality line. Volcano plots are shown on the right. See text for more interpretation.

Indeed, estimating the variance based on two individuals is like estimating the average based on one. Having this metaphor in mind helps to understand the issue of estimating gene variance with only three samples per group.

Variance structure affects performance of methods. The variance structure is also important to consider when choosing a certain method to test for differential expression. Heterogeneity of variance is such a situation. If the groups are not equally variable, one should be careful with the results obtained from a model that assumes that variances are equal for the two groups.

Which statistic to choose for ranking genes. There are basically three main choices: fold change, ordinary t-test or modified t-test.

Ranking the genes on **fold change** should only be done if one prefers to focus on the absolute differences between groups, thereby deliberately ignoring the information of the variation within the groups. As not all genes have the same variation across the replicates within a group, it is generally better to incorporate this information. Indeed, a gene with high fold change may just be highly variable and thus with low significance of the difference (see Figure 5.12).

An **ordinary t-test** compares the difference between groups with the variation within groups (see Figure 5.13), and is therefore a better alternative. Ordinary t-statistics are not ideal because small variances often result in large t-statistics even if the difference is very small. As the t-statistic fails due to small variances and the fold change due to large variances, their shortcomings are opposite[143] and a compromise is needed.

A **modified t-statistic** is a suitable compromise between the ordinary t-statistic and the fold change. These approaches modify the estimate of the gene-specific variance in order to make them more stable and to avoid them to become too small. This modification is either ad hoc by adding a positive constant (the penalized t), or by estimating the variance using empirical Bayes. The latter, the moderated t-statistic, is preferred as it is statistically better defined and can easily handle more complex experimental designs[153]. The moderated t-statistic has been shown to perform very well compared to most alternatives across a range of sample sizes in various real-life datasets[146],[145]. Jeffery et al.[145] also neatly showed the change in similarity of moderated t to other methods when going from large to small sample sizes. Although there were no large dissimilarities between all methods for the large sample-sized studies, the moderated t was more similar to the ordinary t compared to the penalized t. In small sample-sized studies, in contrast, the moderated t was much more similar to the penalized t than to the ordinary t. The reason is that large sample-sized studies allow a proper estimation of the variance, so that the moderated effect weakens. The real issue of the instability of the variance estimation occurs in studies with small

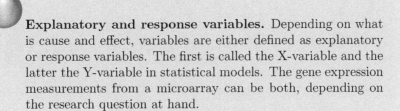

Explanatory and response variables. Depending on what is cause and effect, variables are either defined as explanatory or response variables. The first is called the X-variable and the latter the Y-variable in statistical models. The gene expression measurements from a microarray can be both, depending on the research question at hand.

StatsBox 5.14: Explanatory and response variables

sample sizes. Both the moderated t and the penalized t deal with small variance genes by using information across all genes. This common rationale of the two diffrent approaches results in quite similar conclusions, particularly when compared to the other methods for determining differential expression.

Which statistic to choose for ranking genes.
The moderated t-statistic[153] is recommended as it is statistically well defined, can easily handle complex experimental designs, and has been shown to perform very well compared to most alternatives across a range of sample sizes in various real-life datasets[146],[145].

5.5.2.5 Linear models

The general linear model (GLM) underlies most of the statistical analyses used in empirical research. It is the foundation for the t-test, analysis of variance (ANOVA), analysis of covariance (ANCOVA), regression analysis, and many of the multivariate methods including factor analysis, cluster analysis, multidimensional scaling, discriminant function analysis, canonical correlation, and others[159]. The GLM allows us to summarize a wide variety of research outcomes, like comparing multiple groups or testing for interactions between the levels of two (or more) variables. Because linear models offer various possibilities to model combinations of parameters, it is always important that the researcher carefully considers how to specify the general linear model.

In the previous sections we explained the concepts of hypothesis testing through the use of the t-test because of its simplicity. These concepts are however not limited to t-tests, but naturally extend to more complex models like ANOVA and regression models.

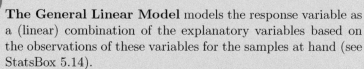

The General Linear Model models the response variable as a (linear) combination of the explanatory variables based on the observations of these variables for the samples at hand (see StatsBox 5.14).

More specifically, the response variable y is modeled as a function of the p explanatory variables x_j ($j = 1, \ldots, p$) and an error term ε. The explanatory variables x_j can be continuous (i.e., regression) or categorical (i.e., ANOVA) or a combination of both.

$$y_i = \beta_0 + \sum_{j=1}^{p} \beta_j \cdot x_{ij} + \varepsilon_i$$

The error term ε is representing the unexplained variation in the response variable, and the coefficients of the explanatory variables β_j act as the "weight" for each variable. The model will estimate an intercept β_0 and a coefficient β_j for each gene i in the model.

Linear models allow to calculate the significance of each of the coefficients. The null hypothesis is that the coefficient is zero, implying that there is no relationship between the explanatory and the response variable.

StatsBox 5.15: The general linear model

5.5.2.5.1 Fitting the model The model (i.e., the combination of the explanatory variables) is constructed based on the observations of the dataset at hand (see StatsBox 5.15). This is done by estimating the coefficients β_0 and β_j so as to give a "best fit" of the data. Most commonly the best fit is evaluated by using the least squares method, but other criteria have also been used. The least squares method, as already suggested by its name, tries to minimize the residual sum of squared errors (RSS). Indeed, the smaller this unexplained variation by the model, the better the model fits the data. The RSS is simply the squared difference of the true response with the response fitted by the model (see StatsBox 5.10 for more information on sum of squares).

$$RSS = \sum_{i=1}^{n} (y_i - \beta_0 - \sum_{j=1}^{p} \beta_j \cdot x_{ij})^2 = ||\beta X - y||_2^2$$

The latter is a more elegant formula of the former, where β_0 has been put in the vector β, and where X gets an extra column with all 1 values. It can be shown that the least square (LS) method yields *unbiased* estimates of the true parameters β_0 and β_j.

The residual sum of squared errors is the typical example of a loss function. A loss function is a function that is minimized in the process of fitting a model, and it represents a selected measure of the discrepancy between the observed data and data "predicted" by the fitted model.

Model building Not all explanatory variables may be important to predict the response variable. When there are several explanatory variables, researchers generally want to know which variables are important and which are redundant. The search for the best combinations of explanatory variables is called "model building" or "variable selection." There are many different ways to select the most important variables.

Typical approaches are stepwise procedures such as forward selection or backward selection. In stepwise forward selection, explanatory variables are added to the model one by one. The first step evaluates all of the variables, and the most significant variable is entered first. Then on each new step, the variable with the smallest p-value in combination with the first selected variable is selected, and the selection of additional variables stops when none is significant. Backward selection proceeds in the opposite way. The procedure starts with all variables in the model, and then removes variables one at a time base on the largest p-value. Despite the appealing logic of their rationale, these approaches are rarely used. A better alternative is all subsets regression, which selects the best combination of explanatory variables by fitting all possible combinations. There is a wide variety of criteria for choosing the best possible model. is used. These criteria are based on the goodness-of-fit of the model (see StatsBox 5.2), but take also the number of explanatory variables into account. This is important, as models with more variables always provide a better fit (see overfitting in Section 5.6.2.1). Nowadays, LASSO (described in Section 5.5.2.5.4 and illustrated in Section 5.6.4.3) provides a more interesting alternative.

Use model for prediction After building the model, the response of a gene i can be predicted based on model from StatsBox 5.10. The uncertainty, or reliability, of the prediction can be assessed by looking at the confidence or prediction intervals of the estimates of coefficients. Confidence intervals always accompany the estimates of the coefficients (see StatsBox 5.10), and prediction intervals can easily be estimated. Whereas confidence intervals are used for the true population estimates, prediction intervals are used to quantify the distribution of the prediction of individual points. As prediction intervals thereby incorporate extra subject-specific uncertainty, they are wider than confidence intervals. Actually, they are typically so wide they are rarely used.

Example. Imagine that one wants to model the relationship between gene expression and age and gender based on gene expression measurements of 200 subjects (100 males and 100 females, with ages ranging from 12 to 85 years). Say that for a specific gene ABC the model estimated the intercept to be 7, the gender coefficients β_{gender} to be 3, β_{age} to be 5 and $\beta_{gender \times age}$ to be -2, and that all these estimates were important (i.e., their p-values are significant). The fitted model would accordingly look like

$$y_{ABC} = -7 + 3 \cdot x_{gender} + 0.5 \cdot x_{age} - 0.2 \cdot x_{gender \times age}$$

As gender is a categorical variable with two levels (male and female), the variable x_{gender} is either 0 (females) or 1 (males). This means that females have an average expression of gene ABC of $-7 + 0.5 \cdot age$ and males of $-4 + 0.3 \cdot age$. A female of 40 years old is therefore predicted to have an average expression of ABC of 13 ($-7 + 20$), and a male of 30 years of 5 ($-4 + 9$). The β_{age} coefficient of 0.5 indicates that the expression level for females increases on average with 0.5 per year. By the way, the term $\beta_{gender \times age}$ is actually an interaction term (see StatsBox 5.17) between a categorical and a continuous variable.

5.5.2.5.2   ANOVA   An ANOVA, the abbreviation for analysis of variance, analyzes the differences in expression levels between two or more groups. It is a linear model (see StatsBox 5.15) in which all explanatory variables are categorical (see StatsBox 5.16). The response variable needs to be a numerical, continuous variable and is in microarray typically the expression levels of a single gene.

An ANOVA basically partitions the observed variation in gene expression between the samples into components due to the different explanatory variables and unexplained variation (the residual noise). It determines the significance of each of the explanatory variables by comparing the differences between the groups to the variation within the groups (similar to the *t*-test, see Section 5.5.2.2.1).

Whereas a t-test is limited to the comparison of two levels (e.g., treatment A and control) of a single variable (here treatment), ANOVA can handle more levels per variable (like treatment A, treatment B and control), and/or more variables (like treatment A and control, and time: 2 hours and 24 hours after treatment). This allows to explore whether certain effects are condition-dependent, i.e., to test whether there is a significant interaction between certain variables (see StatsBox 5.17).

Interactions   can come in different ways. When faced with a significant interaction between two explanatory variables, it is therefore wise to observe the behavior in intensity plots. Let's take a hypothetical microarray experiment with two explanatory variables: knockout status (knockout vs. wildtype) and treatment (control vs. treatment). Figure 5.19 shows four possible effects

Scale of measurement. Data can be either in a categorical or a continuous scale. Continuous variables are numeric values that can be ordered sequentially, and that do not naturally fall into discrete ranges. Examples are gene expression levels, age and weight. Categorical data are data that fall in a limited number of discrete classes. Examples are gender, treatment and disease status. The scale type of the response and the explanatory variables is one of the major criteria in the choice of an appropriate statistical model, as shown in the simplified table below.

| | | Explanatory Variable | |
		Categorical	Continuous
Response Variable	Categorical	Chi Square Test	t-test, ANOVA
	Continuous	Classification	Regression

Categorical scales can be further subdivided in nominal (when the scales are unordered) or ordinal scale (when the scales have a certain ordering).

StatsBox 5.16: Scale of measurement

Interaction. An effect of interaction occurs when the relation between two variables is modified by another variable. In a statistical model, this implies that the effect of one explanatory variable on the response will depend on the values of another explanatory variable. If there is no interaction between two variables, it is said that their effects on the response are *additive*.

See text for an example of an interaction between a categorical and a continuous variable (i.e., gender and age) and of an interaction between two categorical variables (knockout and treatment).

StatsBox 5.17: Interaction

on gene expression. In general, if the dotted lines are parallel, there is an *additive* effect. If the dotted lines have significantly different slopes, there is a significant *interaction*.

What to do with main effects when there is an interaction?
1. It is not allowed to exclude (non-significant) main effects from the model if their interaction is significant.
2. It is dangerous to interpret main effects when there are interaction effects. The interaction namely indicates that there is no overall effect, but rather an effect that changes with the levels of another variable.

1. Treatment and knockout effects, but no interaction:
   There is up-regulation in the knockout and there is up-regulation by the treatment. However, the treatment effect is similar for knockout and wildtype mice. Hence, the effects of knockout and treatment are not interactive but additive.

2. No main treatment or knockout effects, but an interaction:
   The treatment has a positive effect in knockout mice but a negative one in wildtype. When pooling the knockout and wildtype mice together, the treatment effects would cancel each other out. This is why the *main* treatment effect is not significant. One could also describe the pattern

from the perspective of the knockout effect. There is no knockout effect in control conditions, but there is down-regulation in knockout mice after treatment. When pooling the control and treated mice together, a knockout effect would be absent. The *main* knockout effect would consequently not be significant.

3. A main knockout and interaction effect, but no main treatment effect: While there is no treatment effect in wildtype mice, there is an up-regulation by treatment in the knockout mice. Overall, the gene is over-expressed in knockout mice.

4. A main treatment and interaction effect, but no main knockout effect: While the treatment is up-regulating gene expression in both knockout and wildtype mice, the degree of up-regulation differs. The up-regulation is more profound in knockout mice compared to wildtype mice. Interaction effects are often more subtle, as in this scenario.

Example. Let's use the GLUCO dataset, where chondrocytes were treated with glucosamine and IL1-β, together with their respective controls, in every different combination (see the columns of the heatmap in Figure 6.1 for an overview). Differential expression was tested for the glucosamine effect, for the IL1-β effect and for their interaction using a moderated t-test (Limma[156], see Section 5.5.2.2.3). There are consequently three hypothesis tests: (1) whether genes are significantly differentially expressed by glucosamine treatment, (2) whether genes are significantly differentially expressed by IL1-β treatment, and (3) the interaction or whether there are genes where the effect of glucosamine treatment depends on whether the samples received an IL1-β treatment or not.

Figure 5.20 shows four genes that have different expression profiles:

1. Probeset 1368146_at: A gene that is up-regulated by glucosamine treatment. There is no IL1-β effect, and the glucosamine effect is similar for samples that received IL1-β or did not.

2. Probeset 1369773_at: Glucosamine treatment does not change gene expression in samples that received IL1-β. However, in samples that did not receive IL1-β, the gene is down-regulated by glucosamine. This is probably due to the fact that IL1-β treatment already down-regulated the gene in the absence of glucosamine. Note that this might be a false positive finding as the absolute expression values are quite low, but here we only use it for illustrative purposes.

3. Probeset 1367509_at: The expression of this gene is down-regulated by glucosamine. The down-regulation in IL1-β challenged samples is, however, weaker, probably because of the same reason as with the previous gene: IL1-β treatment already down-regulated the gene in the absence of glucosamine.

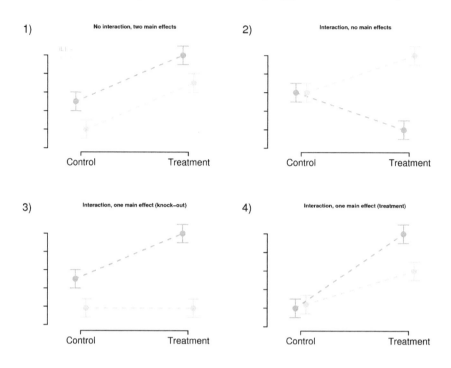

FIGURE 5.19: Scenarios of various interaction effects. The hypothetical dataset is an experiment where a certain treatment, together with a control, has been administered to wildtype and knockout mice. See text for more details.

4. Probeset 1370117_at: There are no significant effects. It seems that IL1-β treatment is somewhat up-regulating gene expression, but because this is mainly driven by two outliers it is not statistically significant.

5.5.2.5.3 Regression Multiple regression models relationships between a response variable and of one or more explanatory variables (see StatsBox 5.15). All these variables are numerical and continuous.

Regression models allow to calculate the significance of each of the coefficients of the explanatory variables. The null hypothesis is that the coefficient is zero, implying a horizontal relationship between the explanatory and the response variable. In other words, increases or decreases in the explanatory variable are not accompanied with a change in the response. Significant p-values indicate that the slope of the relationship between the explanatory and the response variable is significantly different from zero, i.e., that there is a relationship. It implies that the obtained coefficient is unlikely to be due to

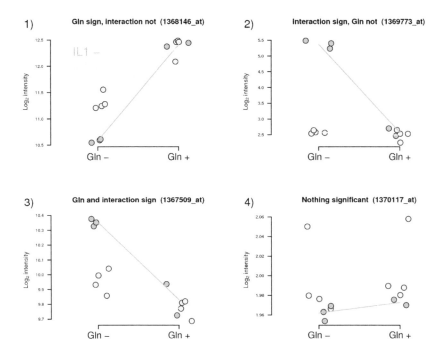

FIGURE 5.20: Four genes with different expression patterns in the GLUCO data to illustrate interaction effects. See text for more details.

chance, indicating that the respective variable appears to be correlated with the response variable.

5.5.2.5.4 Penalized linear models Regression is very sensitive to correlations between explanatory variables. When this happens, the model parameters become unstable (i.e., their estimates have large variances) and can therefore no longer be interpreted.

The instability of minimizing the residual sum of squares (RSS, see Section 5.5.2.5) can be illustrated with an example. Consider the case with two genes k and l that are highly correlated. This dependency can cause the two variables to cancel each other out in the same model. For instance, the estimate of β_k can grow very large when β_l is growing appropriately large in the negative direction. Therefore, the resulting models can produce wildly differing β_j values (and subsequent predictions) when other samples were used, or when the measurements of a sample would have been slightly different. Thus, the final result is that the estimates of β_k and β_l tend to be highly variable and unrealistic. Such high variance models typically become increasingly unrealistic as the correlation between the explanatory variables increases.

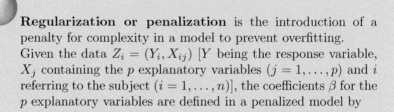

Regularization or penalization is the introduction of a penalty for complexity in a model to prevent overfitting.
Given the data $Z_i = (Y_i, X_{ij})$ [Y being the response variable, X_j containing the p explanatory variables ($j = 1, \ldots, p$) and i referring to the subject ($i = 1, \ldots, n$)], the coefficients β for the p explanatory variables are defined in a penalized model by

$$\min_{\beta} \sum_{i=1}^{n} \{L(Z_i; \beta) + \lambda T(\beta)\}$$

where $L(Z_i; \beta)$ is a loss function (e.g., *RSS*, see Section 5.5.2.5), $T(\beta)$ a penalty function and λ a tuning parameter.
Typical examples of $T(\beta)$ are L_1- and L_2-norm penalizations that can be found in penalized regression methods and support vector machines.
From a Bayesian point-of-view, many regularization techniques correspond to imposing certain prior distributions on model parameters.

StatsBox 5.18: Regularization or penalization

The penalization is typically done by introducing a penalty factor in the least square (LS) method (see StatsBox 5.18). This addition will make the parameter estimates somewhat biased but less variable. This illustrates the concept of the "bias-variance trade-off;" a biased estimator may outperform an unbiased estimator as long as its variance is much smaller.

In microarray data, the explanatory variables are regularly genes. As typically many genes are strongly correlated, the predictions using penalized regression will be more accurate compared to ordinary regression models.

Regularization can be safely used in combination with proper cross-validation for the purpose of prediction, but one should avoid any (biological) interpretation of the obtained set of explanatory variables and their regression coefficients[160].

L_2 regularization L_2 regularization[161],[162] circumvents the problem of strong correlations between the explanatory variables. It gives up the least squares (LS) as a method for estimating the parameters of the model, and focuses instead of the covariance matrix between the different explanatory variables ($X^T X$).

L_2 regularization adds a positive constant to the diagonals of $X^T X$, so that the RSS gets a penalty term proportional to the squared values of β_j[163]. In a formula, it looks like

$$\min_{\beta}\{RSS + \lambda \sum_{j=1}^{p} \beta_j^2\} = \min_{\beta}\{||\beta X - y||_2^2 + \lambda ||\beta||_{L2}^2\}$$

So large β_j will be penalized more. This makes sense in terms of prediction, as large β_j values cause large variations in $\sum_{j=1}^{p} \beta_j.x_{ij}$ when applying the model to new subjects, thereby decreasing the predictive performance of the model.

Ridge Regression[161] is the best-known application of L_2 regularization.

L_1 regularization   While L_2 regularization is very effective in achieving stability and in increasing predictive performance, it does not address another problem with LS estimates, namely the difficulty to identify which of the explanatory variables are the important ones. L_2 regularization bounds the size of the coefficients β_j, but does not set the unimportant β_j to zero. L_1 regularization shrinks unimportant β_j and sets some of them to zero, performing parameter estimation and variable selection at the same time. Contrary to the L_2 penalization, which uses all explanatory variables (genes) in the prediction, only some genes are used in the prediction with the L_1 penalization. Because these sparse models (i.e., models containing many coefficients equal to zero) include only few explanatory variables, they are also easier to interpret[164]. L_1 regularization therefore will often outperform L_2 regularization when there are many irrelevant explanatory variables.

$$\min_{\beta}\{RSS + \lambda \sum_{j=1}^{p} |\beta_j|\} = \min_{\beta}\{||\beta X - y||_2^2 + \lambda ||\beta||_{L1}\}$$

Tibshirani[165] and Efron et al.[166] presented and popularized L_1 regularization under the names least absolute selection and shrinkage operator (LASSO) and by least angle regression (LARS).

L_1 and L_2 regularization   The threshold gradient descent (TGD)[167] and elastic net[168] allow a compromise between the L_1 and L_2 penalizations. Through the choice of a defined threshold, it approximates the L_2 (low threshold) and L_1 (high threshold) penalized estimations.

$$\min_{\beta}\{RSS + \lambda P_\alpha(\beta)\}$$

where

$$P_\alpha(\beta) = \sum_{j=1}^{p}\left[\frac{1}{2}(1-\alpha)\beta_j^2 + \alpha|\beta_j|\right] = \frac{1}{2}(1-\alpha)||\beta||_{L2}^2 + \alpha|\beta|_{L1}$$

Combining the L_1 and L_2 penalties logically results in an intermediate behavior, with fewer regression coefficients set to zero than in a pure L_1 setting, and more shrinkage of the other coefficients[169].

5.5.2.5.5   Mixed models   Mixed models are models that model both fixed effects and random effects. Fixed effects are explanatory variables which levels have been specifically chosen or measured when designing the experiment. Random effects are effects where the levels of the effects are assumed to be randomly selected from an infinite population of possible levels. While the obtained estimates of fixed effects are average differences (in ANOVA) or slopes (in regression), estimates of random effects are variance components.

For the reason of clarity, we restrict examples in this book to fixed effects. Fixed effects are also much more widespread compared to random effects. Continuous measurements (like age) are always fixed as these measurements are specifically obtained from all the subjects. Gender effects are also always fixed. An infinite number of possible levels is impossible as there are only two levels (male and female).

Every linear model contains at least one random effect, namely the subject effect. Indeed, the subjects included in the study should have been randomly selected, and are used to generalize the conclusions to other subjects from the same target population (see Section 5.1). The variance component estimating the random subject effect is the so-called residual variance, the variation between the subjects that is not explained by the model.

Random effects are typically used when there is a dependency in the data in an effect that is not of direct interest. A typical example is when the same subjects have been measured before and after a treatment. First, every statistical model assumes that the samples have been independently and randomly collected. Not acknowledging for the fact that some observations belong to the same subject will lead to flawed conclusions. Second, including subject as a random effect will increase the power to find a significant treatment effect. By modeling the subject-specific variation, the treatment effects are examined within individuals so that more subtle effects are easier to detect.

Mixed models differ from fixed effect models only in the way in which fixed effects are tested for significance[170]. In fixed effect models, the variables are tested using the mean squared residual as the error term. In mixed models, tests use the relevant error terms based on the co-variation of random sources of variation[171]. In the example of the repeated measures per subject, this would be based on how strongly the before and after measurements are correlated within subjects.

5.5.2.5.6   Moderated linear models   Just like a t-test being a specific case of an ANOVA, the moderated t-statistic can be seen as a special case of a "moderated ANOVA" for comparing multiple groups. The moderated F-statistic combines the t-statistics for all the group-to-group contrasts for each gene into an overall test of significance for that gene. The moderated F-statistic tests whether any of the contrasts are non-zero for that gene, i.e., whether that gene is differentially expressed on any contrast. The moderated-F has numerator degrees of freedom equal to the number of contrasts and its denominator degrees of freedom are the same as the moderated-t. The denominator degrees of freedom is the only difference compared to the ordinary F-statistic. The reason is the same as with the moderated t-statistic (see Section 5.5.2.2.3); the added degrees of freedom are due to the fact that the denominator mean squares are moderated across genes[153].

In a complex experiment with many contrasts, it may be desirable to select genes first on the basis of their moderated F-statistic, and subsequently to decide which of the individual contrasts are significant for those genes. This cuts down on the number of tests which need to be conducted and therefore on the amount of adjustment for multiple testing[153].

5.5.3   Correction for multiple testing

A test for differential expression is a statistical test that is repeated as many times as there are genes in the microarray dataset. As shown below, this results in false positives, i.e., genes that are found to be statistically different between conditions, but are not in reality. Multiple testing corrections adjust p-values to quantify and correct for this occurrence of false positives due to multiple testing.

5.5.3.1   The problem of multiple testing

The problem of multiple testing (also referred to as "multiplicity") is the problem of having an increased number of false positive results because the same hypothesis is tested multiple times. It is essential to understand the impact of the multiple testing problem as it explains why microarray experiments often contain false positive findings, particularly when having few replicates and many genes.

The multiplicity problem has its root in the significance level setting (see StatsBox 5.12 for more information, including the relation between significance level and type I error, which is the chance of obtaining false positive findings). Significance levels are typically set to 0.05, so that the researcher is quite confident (i.e., 95% confidence) that the drawn conclusions are correct. This implies that one allows a 5% chance of having a false positive result when testing differential expression *of a single gene*. This allowance for some uncertainty on the made conclusions is unfortunate but inevitable. Any biological experiment will contain some unknown variation, and this prevents us to make universal generalizations when using only a limited sample of the entire population. Even a critical researcher is therefore not inclined to request 100% confidence; 95% confidence is typically seen as "significant enough." This 95%

confidence implies that, even if there is no difference in reality, this difference would be called significant 5 times per 100 experiments or tests.

This concept is rather difficult to grasp, but is easy to illustrate empirically in a computer lab. First, one creates an artificial dataset with 40 columns (the samples) and 1,000 rows (the genes) containing completely random values. Second, a t-test is applied on each row that compares the first 20 with the other 20 samples. As the researcher created the data himself to be random, he knows that every gene that will be called significant is a false positive finding. However, he will observe that around 50 genes or 5% will have a p-value below 0.05.

Another way to understand the logical nature of the multiplicity problem is to imagine an urn with 20 balls of which 19 are black and 1 is white. The chance of randomly sampling the white ball is 1 out of 20 (5%). However, when sampling a single ball (and placing it back in the urn) 50 times, you can expect a much higher chance to sample the white ball at least once. Indeed, there would be a 92.3% chance to sample the white ball $(100 \times (1 - (\frac{19}{20})^{50}) = 92.3)$. This is exactly what happens when testing several thousand genes at the same time; imagine that the white ball is a false positive gene: the chance that false positives are going to be sampled is higher the more genes are tested.

Note that this 5%, the probability of false positive findings, is the significance level that is being controlled by the user, allowing the user to have much control over this expected proportion (see StatsBox 5.12).

5.5.3.2   Multiplicity correction procedures

Multiple testing correction is necessary to provide the researcher with a framework in which he can put the obtained levels of significance. It will generate, in addition to the p-values, adjusted p-values that take the number of tests into account. When analyzing a microarray dataset, a p-value of 0.001 might be a false positive finding just because more than 1,000 tests have been carried out. An adjusted p-value of the same gene of 0.2 underlines the need to realize that this significant p-value might also have been observed just by chance. In other words, multiple testing correction adjusts the individual p-value for each gene to control the overall family-wise error rate (FWER) or the overall false discovery rate (FDR).

There are many different procedures to correct for multiple testing (see Figure 5.21). The most important variation in these methods is how stringently they correct for the number of applied tests. This stringency is a double-edged sword because of the existing trade-off between sensitivity and specificity (see StatsBox 5.19). A stringent correction will allow few false positive genes (non-differentially expressed genes called significant) but may have a high rate of false negatives (differentially expressed genes called non-significant). As will be discussed below, generally less stringent correction methods are chosen in (exploratory) microarray experiments, as a false discovery is preferred above a missed discovery.

Sensitivity and specificity are measures to assess the correctness of a decision between two options. The *sensitivity* or recall rate measures the proportion of successfully identifying a real effect (i.e., the percentage of diseased people diagnosed as diseased), and the *specificity* measures the proportion of successfully rejecting a false effect (i.e., the percentage of healthy people not diagnosed as diseased).

Sensitivity and specificity are inversely related, as diagnosing everybody as diseased will increase the sensitivity at the cost of a decreasing specificity.

See StatsBox 5.12 for the close link of sensitivity and specificity with the occurrence of false positives and false negatives.

StatsBox 5.19: Sensitivity and specificity

Most correction methods can be classified in two main families: correction methods that control either the FWER or the FDR.

5.5.3.2.1   Family-wise error rate   The family-wise error rate (FWER) is the probability of making false discoveries, or type I errors (see StatsBox 5.12), among all the hypotheses when performing multiple tests. FWER based correction methods correct the p-value for each gene so that this overall error rate is controlled. There is a variety of methods to control the FWER, but most of them can be categorized into one of the following three types of methods: single-step, step-down or step-up procedures.

Single-step procedure   In a single-step procedure, the multiplicity adjustment is the same for all genes, regardless of the ordering of their unadjusted p-values. The most widely used single-step procedure is the Bonferroni method. It is a very logical and simple, but a very stringent correction method. Basically, the p-value of each gene is multiplied by the number of genes in the gene list. If the corrected p-value is still below a user-specified error rate, the gene is called "significant after multiple testing correction."

Let $p_1, ..., p_k$ be the p-values for the in total m genes. Every i^{th} gene is adjusted by simply multiplying its p_i with m (see Table 5.1). As the p-values are multiplied by a positive number, they become larger and hence less significant. Note that one could also divide the significance level α by k, and compare p_i to this adjusted significance level.

$$p_i^{Bf} = p_i \cdot m, \; with \; i = 1, 2, \ldots, m$$

The Bonferroni method regards genes as being independent. This simplification is evidently not true for microarray data that typically contain many correlated genes. The (single-step) Westfall and Young permutation method incorporates this correlation between genes, by permuting all the genes at the same time. P-values are calculated for each gene based on the original dataset and on many permutated datasets. The proportion of permutated datasets where the minimum p-value is less than the original p-value is the adjusted p-value. Note that the permutations make the method relatively slow.

Step-down procedure   A single-step procedure is typically too conservative (the true FWER is smaller than estimated by the Bonferroni method) and therefore lacks power. Step-down procedures improve the power while still preserving strong control of the FWER.

The most well-known procedure is the Holm correction[172], also called Bonferroni-Holm correction, which is a slight adaptation to the Bonferroni method. In the Holm method, only the p-value of the most significant gene needs to be corrected for the total number of genes. Once this gene has been corrected for, it is out of the ball game and can be ignored when correcting the other genes for multiplicity.

The Holm procedure is called a stepwise or sequential method because it adjusts the genes differently. It first orders the genes from most to least significant, and subsequently adjusts each gene in this ordered sequence with a stepwise decreasing correction factor. The Holm procedure to calculate the adjusted p-values is shown in Table 5.1. Note that in the final step p_m is not multiplied because $(m - (m-1)) = 1$. This corresponds with the underlying rationale that there is only a single gene left in the pool so that no multiplicity issue arises.

Just like Bonferroni, the Holm method regards genes as being independent. The (step-down) Westfall and Young method incorporates this correlation between genes, by permuting all the genes at the same time. The step-down Westfall and Young follows a step-down procedure similar to the Holm method, combined with a bootstrapping method to compute the p-value null distribution.

Step-up procedure   The Hochberg method (1988) is very similar to the Bonferroni-Holm method, but orders the p-values from large to small. It is consequently more powerful and reduces the number of false negatives. This is however still an FWER method, and should not be confused with the Benjamini-Hochberg method which is an FDR method (see Section 5.5.3.2.2).

5.5.3.2.2   False discovery rate   It has been shown that traditional approaches to control the FWER are too conservative when applied to microarray data. Recent attention has been focused on the control of false discovery rate (FDR). Where the FWER controls the overall probability of having false

Multiple testing correction procedures.
Every multiple testing correction procedure uses some error rate to measure the occurrence of incorrect conclusions. The focus is generally on errors due to false positive conclusions (the non-differentially expressed genes called significant) while keeping the number of false negatives (the differentially expressed genes not called significant) as small as possible.

Three main types of error rates are currently being used:

FWER: the expected occurrence of false positives V among **all tested genes** G, assuming that no single gene is differentially expressed.

$$FWER = E(\frac{V}{G})$$

FDR: the expected occurrence of false positives V among **the genes called significant** R, assuming that no single gene is differentially expressed.

$$FDR = E(\frac{V}{R}) \; if \; R > 0 \; and \; 0 \; otherwise$$

pFDR: the expected occurrence of false positives V among **the genes called significant** R, allowing a proportion of the genes to be differentially expressed.

$$FDR = E\left[(\frac{V}{R}) \mid R > 0\right]$$

StatsBox 5.20: Multiple testing correction procedures

significant tests, the FDR controls the probability of having false tests among all the significant tests (see StatsBox 5.20).

An FDR method estimates the proportion of false positive findings amongst the genes that were selected to be differentially expressed. The idea is to allow and define a number of false positives in the genes that are selected rather than looking at the number of false positives in all of the genes. FDR based correction methods correct the p-value for each gene so that this overall false positive rate is controlled (see StatsBox 5.20). The FDR methods[173],[174] control the expected FDR to be less than a given value. The term "control" means that they provide an upper bound rather than an estimate. The term "expected" means that they do not provide a specific control of the FDR for a given dataset, but rather a generally valid control.

FDR correction is much less stringent compared to FWER methods. At a price of tolerating more false positives, this method will result in less false negative genes. While FWER methods only allow very few occurrences of false positives, FDR methods allow a percentage of positives to be false positives. Similarly as with the Holm correction, the FDR correction becomes more stringent as the p-value decreases (see Table 5.1). Note that the resulting adjusted p-values should keep the same ranking as the raw p-values. This is simply done by setting *adjusted* $p_{(i+1)}$ equal to *adjusted* $p_{(i)}$ in the situation where *adjusted* $p_{(i+1)} <$ *adjusted* $p_{(i)}$.

The different philosophy behind FDR has an implication on how to interpret the adjusted p-values. An FWER of 0.05 indicates that 5% of *all* the tested genes may be significant just by chance. An FDR of 0.05 indicates that 5% of *only the significant* genes (after correction) may have been identified by chance (i.e., the false positives).

Various procedures to control the FDR have been proposed[173],[174],[175] (see Pounds et al.[176] for a review), but the original FDR method of Benjamini and Hochberg[173] is still the most popular. Unlike the the Holm method, which is a step-down method starting from the smallest p-value $p_{(1)}$, the Benjamini-Hochberg method is a step-up method starting from the largest p-value $p_{(k)}$. The Benjamini and Hochberg FDR is calculated as shown below (see also Table 5.1).

$$p_i^{BH} = p_i \frac{m}{order(p_i)}, \ with \ i = 1, 2, \ldots, m$$

The Benjamini and Hochberg method[173] is theoretically only valid when the genes are not correlated, which is untrue in reality. The Benjamini and Yekutieli method[174] is valid for any level of correlation between the genes, but is so conservative that almost no one uses it. Simulations suggest that the Benjamini and Hochberg method[173] is unlikely to fail for realistic scenarios[177], and is therefore widely used as it is not too conservative.

The FDR procedure is particularly interesting because microarray experiments are typically exploratory by nature. The positive findings of these

TABLE 5.1: Calculation of adjusted p-values by the Bonferroni, the Holm and the Benjamini-Hochberg procedure. The adjusted p-values are ranked from small to large. (Note that the resulting adjusted p-values should keep the same ranking as the raw p-values. This is simply done by setting adjusted p(i+1) equal to adjusted p(i) when it was lower. Also, if adjusted p-values exceed 1, they are set to 1.)

Gene	p-value	Bonferroni	Holm	FDR BH
gene A	$p_{(1)}$	$m \cdot p_{(1)}$	$m \cdot p_{(1)}$	$m \cdot p_{(1)}$
gene Z	$p_{(2)}$	$m \cdot p_{(2)}$	$(m-1) \cdot p_{(2)}$	$\frac{m}{2} \cdot p_{(2)}$
gene D	$p_{(3)}$	$m \cdot p_{(3)}$	$(m-2) \cdot p_{(3)}$	$\frac{m}{3} \cdot p_{(3)}$
...	...	...	...	...
gene X	$p_{(i)}$	$m \cdot p_{(i)}$	$(m-i).p_{(i)}$	$\frac{m}{i} \cdot p_{(i)}$
...	...	...	...	...
gene K	$p_{(m-2)}$	$m \cdot p_{(k-2)}$	$3 \cdot p_{(m-2)}$	$\frac{m}{m-2} \cdot p_{(m-2)}$
gene B	$p_{(m-1)}$	$m \cdot p_{(k-1)}$	$2 \cdot p_{(m-1)}$	$\frac{m}{m-1} \cdot p_{(m-1)}$
gene L	$p_{(m)}$	$m.p_{(k)}$	$p_{(m)}$	$p_{(m)}$

hypothesis-generating experiments are typically validated in follow-up studies. A false discovery is therefore not so disastrous as a missed discovery, so that false positives are preferred above false negatives. In this context, the FDR can naturally be translated into terms of the costs of attempting to validate false positive results[176].

5.5.3.2.3 Positive false discovery rate The Benjamini-Hochberg FDR method is still quite conservative as it assumes that all null hypotheses are true, i.e., that none of the genes are differentially expressed. A general expectation is however that at least some of the genes in the microarray experiment are differentially expressed. Storey[178],[179] followed this rationale and implemented the positive false discovery rate (pFDR).

The pFDR procedure consists out of two main building blocks: the estimation of the number of non-differentially expressed genes (m_0), and the estimation of the expected number of false positives. For the estimation of m_1, Storey[179] regards the p-values (of the in total m genes) as a mixture of m_0 p-values of non-differentially expressed genes and m_1 p-values of differentially expressed genes (with $m_0 + m_1 = m$). As discussed in Section 6.1.6 on page 247, the p-values of non-differentially expressed genes are uniformly distributed between 0 and 1. Storey[179] therefore estimates m_0 as two times the number of genes larger than 0.5 (see Section 6.8 on page 248 for a figure and explanation). For the estimation of the expected number of false positives, the sample labels are permuted to calculate the p-values when no genes would be differentially expressed (the so-called "null distribution"). The number of genes from this permuted dataset that is significant for a user-specified significance level can consequently be used as a measure of the expected number of false positives under the null distribution, referred to as $\widehat{E(V_0)}$. The pFDR is then this estimate multiplied by the estimated proportion of non-differentially expressed genes in the microarray dataset.

$$pFDR = \frac{\hat{m}_0}{m} \widehat{E(V_0)}$$

5.5.3.2.4   False negative control   The false negative rate (FNR), or sensitivity, refers to genes that are truly differentially expressed but have not been called significant. As warned by Pawitan et al.[180] researchers need to be aware of the FNR when controlling the FDR. Indeed, one does not want to lose too many of the truly differentially expressed genes by setting the FDR too low. The use of FDR should therefore ideally be accompanied by a FNR assessment. Pawitan et al.[180] provide a way to consider the FNR by computing FNR curves routinely together with FDR curves.

Smet et al.[181] describe a procedure based on receiver-operating characteristic (ROC) curves. This approach can be useful to select a rejection level that balances the number of false positives and negatives and to assess the degree of overlap between the two sets of p-values.

Although valuable in certain situations where the microarray experiment is more confirmatory than exploratory by nature, FNR is generally considered to be less important. Most studies therefore focus on errors due to false positive conclusions while keeping the number of false negatives as small as possible.

5.5.3.3   Comparison of methods

Differential expression is typically determined by comparing p-values (derived using an appropriate hypothesis test) to a certain rejection level. This selection, however, is not possible without accepting some false positives and negatives since in reality the two sets of p-values, associated with the genes whose expression is and is not different, overlap[181]. This is why a decrease in false positives is generally inextricably bound up with an increase in false negatives. This is due to the trade-off between false positives and false negatives, something quite inevitable that goes back to the universal sensitivity-specificity trade-off (see StatsBoxes 5.12 and 5.19). Logically, all multiplicity correction methods can be positioned in this trade-off gradient (see Figure 5.21).

Figure 5.22 shows the unadjusted and adjusted p-values for the ALL dataset when testing between ABL/ALL and NEG. Very importantly, it is clear from the figure that the rankings of the genes remain unchanged when comparing the different methodologies. As could be expected, the two FDR methods and the two FWER methods cluster together. The Bonferroni method is the most and Storey's pFDR the least conservative.

It has become obviously clear that FDR methods are preferred above FWER methods in high-dimensional, exploratory data. This preference for FDR

False positives vs. false negatives

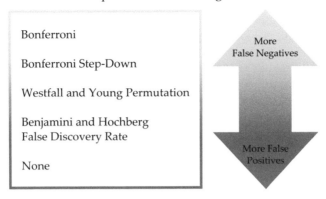

FIGURE 5.21: A schematic table of the most widely used multiplicity correction methods, ranked according to the trade-off between false positives and false negatives.

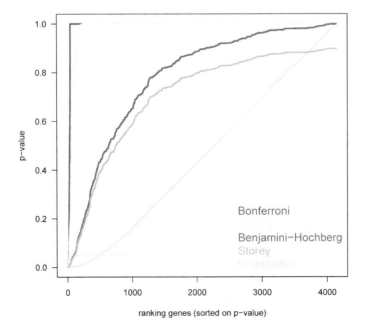

FIGURE 5.22: Adjusted and unadjusted p-values for the ALL dataset.

methods is ubiquitous in microarray gene expression studies, as these generally have exploratory purposes. There is no real consensus yet which FDR method to use. SAM[150] uses for example Storey's pFDR, while Limma[153] has the Benjamini-Hochberg FDR implemented.

As mentioned in StatsBox 5.7, testing for differential expression involves two steps: (1) the ranking of the genes based on a statistic and (2) the arbitrary choice of which number of top-ranked genes are called significant. The first step, the ranking of the genes, is the most important one of the two. Now, multiplicity correction only affects the arbitrary threshold choice, and does not change the rankings of the genes. This is why we focus on the top genes and their unadjusted p-values, and use FDR or pFDR only to position this evidence in the context of the number of tests performed.

Nevertheless, FDR is preferred above FWER methods. Whether pFDR or FDR is chosen is generally not so important in practice. It frequently occurs that, based on the same p-value gene list, FDR results in no single and pFDR in several significant genes. Which of the two multiplicity corrections turns out to be the best will depend on the biological relevance and the unadjusted p-values of the top genes. In practice, we will however always be inclined to check the biological relevance of the top genes if they have quite low unadjusted p-values. In other words, we would still have discovered the genes even when being non-significant after FDR. Moreover, we usually automatically incorporate our knowledge about pFDR and FDR in our interpretation. A finding that is significant after pFDR adjustment will make us less enthusiastic as a significant gene after pFDR adjustment. In conclusion, the choice between FDR and pFDR may be regarded as a personal preference, which is not problematic as it is in itself already based on a subjective concept that needs to be interpreted relatively rather than absolutely.

5.5.3.4 Post-hoc comparisons

So far, we have only considered multiple testing from the perspective that the same test was applied on many genes. There are however also situations where multiple tests are conducted for a single gene. These are typically pairwise comparisons between multiple groups.

When testing for differential expression between more than two groups, one sometimes wants to to test for an overall significance, i.e., whether there is significant differential expression between any of all the studied groups. Such a test is typically done using a (parametric) ANOVA F-test or a (non-parametric) Kruskall-Wallis test. A significant ANOVA result suggests that at least one of all the studied groups is differentially expressed compared to the others. Multiple comparison procedures are then used to determine which group is different from which.

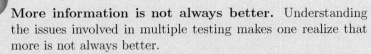

More information is not always better. Understanding the issues involved in multiple testing makes one realize that more is not always better.

Imagine a differentially expressed gene with a p-value of 0.00001. If measured with a focused array containing 100 potentially interesting genes, its (Bonferroni) adjusted p-value is 0.001, which indicates a quite significant difference. If measured with a genome-wide array containing 50,000 genes, the corrected p-value would be 1 and completely non-significant. This is the main reason why it is valuable to exclude the non-informative genes before the analysis (see Section 5.3).

StatsBox 5.21: More information is not always better

5.5.3.4.1 Pairwise comparisons among all groups Comparing k means involves $k(k-1)/2$ pairwise comparisons. There are several correction methods, but the Tukey procedure is a widely used method.

5.5.3.4.2 Pairwise comparisons of all groups vs. one reference group The Dunnet test controls the family-wise error rate when comparing multiple groups to the same group. This is a common situation in compound profiling experiments where all studied compounds are compared vs. a control or vehicle.

5.5.4 Statistical significance vs. biological relevance

Statistical significance does not always imply biological relevance. First, because of multiple testing problems (see Section 5.5.3.1), the top gene lists generally will contain a substantial amount of genes that are called significant just by chance, the so-called false positive results. Not all top-ranked genes are consequently biologically relevant.

Second, there are more characteristics than significance alone to rank the genes. Imagine a comparison of a treatment effect that affects very many genes. If one measures the alterations in gene expression in tens of thousands of transcripts and the experimental design was such that two thousand genes are altered, what are the biologically relevant changes? Are they those that have the highest alteration? Or are they those that show the largest significance in a statistical test? Or are they only those that are altered in a certain pathway that one expects to be affected? It's very difficult, and often impossible, to say.

For sure, it is important not to focus on significance of the difference alone, but to look also at the fold change of the difference and the absolute expression level of the gene[103]. Even with a highly significant p-value, a tiny change in mRNA abundance will most likely be of no interest to the scientist when there are other genes with a slightly worse p-value but a larger change between the treatment groups. The absolute expression level is also of importance, as genes measured at low amounts of expression tend to be less reliable.

Finally, the most significant gene is not necessarily the most important gene. "Significant" means "important" in normal English, while in statistics it means "probably true," indicating that it is likely not due to chance. A significant result is probably true, but not necessarily important. For example, gene expression studies often do not measure the true biological key process but rather other processes that are the downstream effects of this key process. This particularly holds in experiments with acute treatments where the mRNA is collected a certain time after the treatment. The highly significant genes are downstream effects which are indeed very probably true, but are less important than the upstream key processes.

5.5.5 Sample size estimation

The number of subjects to include in a study, the sample size of the study, is an important consideration in the design of experiment. "How many subjects do we need to study?" is indeed a frequent question from experimenters to statisticians. Although the question is a straightforward one, the answer is not. This is due to the need of concrete information as input for sample size estimation. Particularly, one needs to define how large expression differences one would like to be able to detect, and one needs to predict the expected noise levels in the study. If one studies large effects in a population with low biological variation, for instance the effect of a toxic compound in a high dose on a cell culture system on an isogenic background, 4 replicates per group may already be sufficient. If one studies a subtle signal in a heterogeneous population, like the difference in human plasma gene expression between healthy and depressed people, one probably needs more than 100 individuals in each group.

Results of existing comparable studies –or better, a pilot study– are necessary to calculate realistic necessary sample sizes for an experiment. If multiple genes are involved in a single hypothesis, things will be more difficult. In such situations you are likely to be concerned about joint variability and estimates of that are more difficult to obtain.

Microarray studies are often explanatory, making it difficult to claim in advance which type of effects one is expecting. Fortunately, exploratory experiments are primarily intended to generate hypotheses rather than to confirm them. Sample sizes are therefore less relevant. Generally, the relative ranking of the genes according to potential relevance is more important than the absolute decision which genes are significant.

Sample size is closely tied to statistical power, which is the ability of a study to enable detection of a statistically significant difference when there truly is one. But for the same sample size, it will be easier to detect the "truly" differentially expressed genes if their fold changes are larger, if they are with many and if the background variability is smaller.

Concerning the number of replicates in small well-controlled studies, we suggest using always at least 3 replicates. If one replicate is outlying, then you could still presume the other two, who are highly correlated, to be 2 reliable replicates. When faced with two very distinct replicates, it can be difficult or impossible to determine which of the two is the correct one. Of course, outliers should not be removed just like that. There should always be good evidence that their distinct expression profile is because of technical and not biological reasons.

If a researcher decides to continue with 3 replicates, he should however realize that the power increases exponentially with sample size. In other words, adding an extra replicate in a 3+3 design (making it a 4+4 design) results in a big improvement in power, while going from a 30+30 to a 31+31 design will not drastically improve the power.

5.6 Supervised prediction

Microarrays are very useful in the discovery of "gene signatures" that can predict a certain individual's characteristic. Microarrays can for example provide clinicians with valuable biomarkers for disease status or treatment sensitivity.

Extracting molecular signatures from the high-dimensional data is however a difficult and complex task. The main goal is to select a small number of features (genes) with high predictive accuracy[182]. Classification algorithms in combination with feature selection methods serve this purpose. The entire process of class prediction, or classification, generally consists of three main building blocks:

1. Feature selection: select a subset of relevant genes for building the classification model.

2. Classification: predict the group label of individual samples based on a model or a machine learning technique.

3. Performance assessment: estimate the error rates when classifying an independent set of samples.

Supervised prediction is actually broader than just classification. Classification is a specific type of supervised prediction where the response is categorical. Sometimes the response may however be continuous or time-to-event.

TABLE 5.2: Conceptual differences in the approach between classification and hypothesis testing

	Classification	Hypothesis testing
Number of genes used	Multivariate	Univariate
Level of interpretation	Individual	Population
Modeling the gene expression	Response	Explanatory
Rationale	Decision	Exploratory

When microarray experiments are designed to predict compound sensitivity, the response may be IC_{50} measurements which are continuous measurements. When the purpose of the study is to predict survival or time to relapse, the response is time-to-event data. Many statistical models and machine learning techniques commonly used for classification naturally translate to continuous responses.

5.6.1   Classification vs. hypothesis testing

Although classification is a supervised method just like hypothesis testing, it is conceptually and methodologically very different. The 4 most profound differences between classification and hypothesis testing are listed in Table 5.2.

First, gene-by-gene tests for differential expression ignore the multivariate nature of gene expression. By testing one gene at a time one will miss trends or interactions that exist between different genes. Classification and clustering methods are multivariate models, and they respectively exploit and explore how the genes co-vary in expression levels. This latter is in statistical terminology often referred to as the "covariance structure" of the expression data.

Second, classification and hypothesis testing work on a different level. Hypothesis testing works at the population level; it tests for *average* differences between certain groups. Classification does not work on population but on individual level, as it is used to classify individual samples. A gene that can classify people as diseased or healthy can be used as a diagnostic tool, not necessarily a differentially expressed gene. Hypothesis test would check whether the difference in the averages between diseased and healthy groups is significantly different. Even if a gene is *on average* over-expressed in the population of diseased patients, it may still be useless for individual diagnosis if there is a considerable overlap in expression levels between healthy and diseased subjects. For diagnostic purposes, it is not important to have a high overall significance, but to have a high accuracy to classify a single patient.

Third, classification and hypothesis incorporate the gene expression data at a different place in the model. Classification treats gene expression as explanatory variables while hypothesis testing includes them as the response variable (see StatsBox 5.14). In hypothesis testing, one evaluates the effects of disease status (healthy or diseased) on gene expression. Consequently, disease status is the explanatory and gene expression the response variable. In classification one evaluates the effect of (combinations of) gene expression levels on disease status. Here, disease status is the response and gene expression the explanatory variable.

Finally, the philosophy is different between the two approaches. Classification is a predictive analysis in which clear decisions have to be made. Hypothesis testing of course also involves decision making, like whether a difference is significant or not. In general, however, the genes are primarily interpreted on their relative ranks. If some top-ranked genes appear potentially relevant and biologically interesting, they are subsequently validated in a follow-up experiment. In this context, hypothesis testing has more an exploratory nature, and the real decision will only be made in the follow-up study.

5.6.2 Challenges of microarray classification

There are two major challenges in the search for gene signatures based on microarray data:

1. The superabundance of genes leads to overfitting.

2. The strong correlations between genes leads to non-unique solutions and impedes the biological interpretation of the obtained signature.

5.6.2.1 Overfitting

Overfitting is fitting a statistical model that has too many parameters. Even an absurd and false model may fit the data perfectly as long as the model has enough complexity (enough genes) in comparison to the number of samples. Overfitting is harmful as it reduces or destroys the ability of the model to generalize beyond the fitting data. Microarray data contain much more genes than samples (the large p, small n problem; see Section 5.2). This makes that there are many gene combinations possible for accurate classification of the relatively few samples.

Figure 5.23 shows three hypothetical situations that help to understand the reason for overfitting in classification. The upper panel shows two simple two-dimensional problems. In both cases, one searches for a combination of two genes (gene A and B) to discriminate diseased and healthy individuals. The left situation is problematic due to a small overlap in expression values between the two groups. The right situation is problematic because the gap between the two groups is very large. Although the left situation appears to be more problematic, the opposite is true. Indeed, the next step is to classify new patients as diseased or healthy. In the left situation it is easy to predict the disease status of the two new patients. In the right situation, there are so many ways to discriminate the two groups that the classification of the new samples depends on quite arbitrary choices.

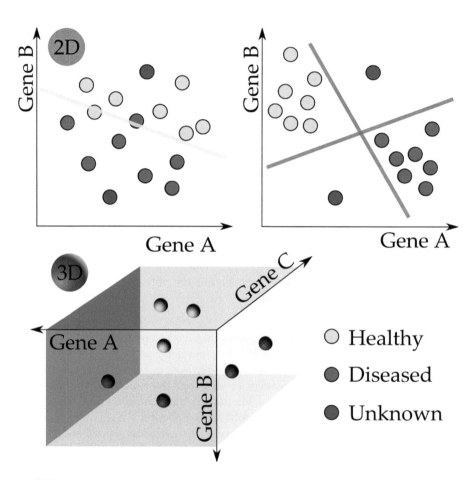

FIGURE 5.23: Three hypothetical situations that help to understand the reason for overfitting in classification. The key message is that large gaps (upper-right graph) or situations with more genes than samples (bottom-graph) are problematic for classification. See text for a more detailed interpretation.

We now have discussed an optimistic scenario with a dataset containing more samples than genes. In reality, we have much more genes than we have samples. Let's imagine two individuals (one diseased and one healthy) and three genes (genes A, B and C) like examplified in the bottom panel. This 3-dimensional plot indicates that there is a huge number of possible combinations to classify disease status, making it very difficult to predict the status of a new incoming patient.

The combination of many genes with small sample sizes is however not the only cause of overfitting. Obtained results may also be specific for the used training samples or for the used selection/classification methods.

Michiels et al.[183] showed that the selection of training samples can influence the results using previously published material. From seven large studies aimed at predicting prognosis of cancer patients by microarrays, they found that the obtained results depended on how the patients were selected in the training set. Because of problems with this aspect of study design, the outcomes of these studies were biased. Although it cannot be verified, a potential explanation is that only positive stories make it to nice publications (the publication bias, see Section 5.8.2). Some earlier or other analyses, based on different training samples, may have shown different results but did not make it to the public domain because of higher misclassification error rates.

The excellent work of Ruschhaupt et al.[184] indicates and illustrates how results may be overfitted by the choice of classification model. A certain gene signature based on a deliberately chosen classification model may fail when using other classification models. This might be because the assumptions of the used model fit the microarray data properties best, but it might also be due to chance. In the latter situation, we ran in the trap of overfitting and will have difficulties to reproduce the results in future studies. Given the potential differences in results between classification models, signatures may depend (highly) on the algorithm used to extract the signature. Therefore, the scientist has to validate how much of the reported discrimination can be attributed to a real biological difference: the scientist needs to disentangle biology and algorithm[184].

5.6.2.2 The bias-variance trade-off

The most intuitive demonstration of overfitting is the bias-variance dilemma[164]. Although statistical models try to fit the data well, a perfect fit is practically impossible. The mismatch between a model and data can be decomposed into two components: bias that represents the approximation error, and variance that represents the estimation error[104] (see StatsBox 4.1). There is a trade-off between bias and variance; if one tries to model the data very accurately (low bias), the model parameters will be less generalizable (high variance) and vice versa.

Now, the link with overfitting (see Section 5.6.2.1) is clear. Using more genes for classification purposes will inevitably result in a better fit of the data

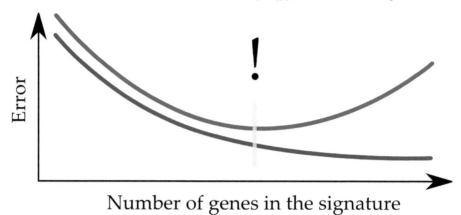

Number of genes in the signature

FIGURE 5.24: The problem of overfitting in classification. The misclassification error is shown in blue for the samples used to train the model, and in red for the samples that test the model. Overfitting occurs when the test error increases while the training error continues to decrease.

at hand (low bias), but whether this fitted model will hold in new samples becomes more questionable (higher variance) when more and more genes come into play. This is shown in Figure 5.24. Increasing the number of genes in the model will always result in a decreased misclassification error of the training samples. This is because these samples are used to select the genes. At a certain moment, all the relevant genes will have been collected. From this moment, all the extra added genes are not relevant but just happen to have expression levels that by chance help to increase the classification accuracy. For the training samples, it does not matter whether it is stochasticity or reality. It will still increase the classification accuracy. For the test samples the classification accuracy will deteriorate when the added genes are just by chance helping the classification of the training samples. As soon as the test error starts to increase, one knows the model is starting to overfit the training samples at the cost of reproducing the results in independent samples. This turning point can consequently be used to assess the optimal number of genes for classification.

5.6.2.3   Cross-validation

Because of the danger of overfitting, it is of fundamental interest to know if prediction would be accurate in new samples. The most logical way to achieve this is to simply split the samples with known class into two groups, and use one part to train the classifier and the other part to validate performance. This practice of partitioning the data into subsets, analyzing a single subset and subsequently using the other subset(s) for confirming and validating the initial analysis is called "cross-validation." Unfortunately, excluding test samples

Nested Loop Cross-Validation

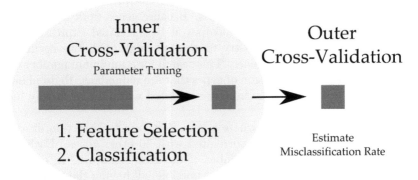

FIGURE 5.25: Scheme of nested loop cross-validation, showing that feature selection and classification are within an outer cross-validation loop and therefore do not see the test samples of that outer CV loop.

from the analysis reduces the sample size of the training set, which is already problematically low in most studies (see Section 5.6.6). As smaller sample sizes reduce the power of the study, the chances to obtain a good classification model will consequently decrease.

More can be done with the dataset than just applying a single cross-validation. One can randomly partition the data more than once, say 50 times, into training and test sets. In each of these 50 cross-validations, a classification model can be built on the training sets and tested on the test sets. Consequently, 50 estimates of misclassification error rates will be obtained based on 50 different test sets. This approach implies that there are basically two cross-validation loops that are nested in each other. The outer cross-validation loop is used to estimate the misclassification rate (like the 50 loops above) and the inner cross-validation loop is used to tune the optimal parameters for a given complete classification procedure (see Figure 5.25). Importantly, the test set used for estimating the misclassification rate is not included in the cross-validation loop for the tuning of the parameters.

5.6.2.4 Non-uniqueness of classification solutions

Non-unique solutions imply that there are many different solutions possible all of which provide a similar answer. In microarray classification, there are typically many gene signatures possible that provide equally accurate predictions but share few common genes[182]. This lack of uniqueness (also called lack of stability) underlines the danger of linking some biological interpretation to genes included in a signature.

The problem of non-uniqueness is due to the presence of many strongly correlated genes in microarray studies. Because genes work in pathways and in gene networks, many genes have strongly correlated expression profiles. Most classification algorithms try to obtain a high predictive accuracy with the smallest possible set of genes. This search for a minimum number of genes generally involves the exclusion of genes that are redundant. Redundant genes are typically genes that are strongly correlated with a gene already included in the signature.

Let's imagine a situation where a gene signature for a certain disease contains a gene that is strongly correlated with 50 other genes. Each of these 50 genes is an equally good member of the gene signature. In other words, there are 50 different gene signatures possible, all having the same predictive power. Now, if only a single gene of these 50 is the real cause for the disease, then the biological interpretation of the signature would be in 49 of the 50 times completely false and misleading.

5.6.3   Feature selection methods

In the context of microarray data, with generally tens of thousands of features (i.e., the genes), signatures will only be useable and interesting when containing a limited selection of genes. Feature selection methods are being used to reduce the high-dimensional microarray data to a more manageable subset that contains the relevant features. Here, the words feature and gene are used interchangeably, as genes are the features in the context of microarray classification.

There is a plethora of methods to select the most important features (see [145],[182] for a review), but there are two main properties that can explain most of the observed differences in the behavior and results of these methods.

1. Univariate vs. multivariate feature selection.

2. Feature selection that is independent vs. dependent of the classification algorithm.

They all have their advantages and disadvantages, and which type of feature selection will result in the best classification performance probably depends on the type of signal in the microarray data.

Univariate or multivariate selection.   A univariate selection will consider each gene separately, thereby ignoring dependencies between the genes. Multivariate selection methods, in contrast, search for interesting combinations of genes.

TABLE 5.3: The top six genes selected by a multivariate selection method (LASSO) and a univariate selection method (PAM).

	LASSO		PAM	
Rank	Gene	Coeff.	Gene	Score
1	ABL1	-2.66	ABL1	0.19
2	TCL1B	-1.84	ABL1	0.16
3	LTF	-1.67	KLF9	0.17
4	FCN1	-0.66	PON2	0.13
5	DSTN	0.60	ABL1	0.12
6	RPS6	0.52	ALDH1A1	0.12
# selected	26		20	

Let's use the comparison of the BCR/ABL group vs. the NEG group of the ALL data to illustrate the difference. More specifically, let's look at the genes selected using a multivariate selection method (LASSO) and using a univariate selection method (PAM). Both LASSO and PAM rank the same gene as most important, namely gene ABL1 (Table 5.3). For PAM, even three of the six probesets measuring ABL1 are in the top five. This makes sense as the main difference between the two groups being compared is a mutation in this particular ABL gene. All the other genes except the first one differ however considerably between the two methods (see Table 5.3).

There is a striking difference between LASSO and PAM in the way that the top genes relate to each other. Figure 5.26 contains some pairwise scatterplots of genes that were ranked high in PAM (see Table 5.3). The first probeset of ABL1 is of course strongly correlated with another probeset also measuring ABL1, but is also considerably correlated with the other genes KLF9 and PON2. This is because they all have been selected on the same criterion, discriminating the BCR/ABL from the NEG samples as much as possible. In other words, PAM selects genes which all discriminate the two groups, resulting in lists of genes with correlated expression profiles.

Figure 5.27 contains some pairwise scatterplots of genes which were ranked high in LASSO (see Table 5.3 and Table 5.4). The expression of ABL is not strongly correlated with the other genes TCL1B, TLF and RPS6. What is also apparent is that these latter three genes are not differentially expressed between BCR/ABL and NEG samples, but they improve *in combination with the ABL1 gene* the discrimination of the two groups (see dashed Grey line in the figures). This illustrates that LASSO selects complementary genes that explain additional variation. Logically, correlated genes are redundant as they explain identical and therefore not additional variation.

There is no universally valid answer for the question whether multivariate or univariate selection performs best. A univariate selection followed by a multivariate classification seems very contradicting. Univariate selection might have excluded the interesting combinations of genes that the subse-

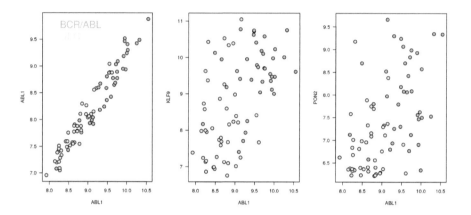

FIGURE 5.26: Some combinations of genes that were ranked high using PAM. The most important gene ABL1 is combined with another probeset of the same gene and with genes KLF9 and PON2, respectively. The samples are colored by the mutation status of the BCR/ABL fusion gene, with (BCR/ABL) and without (NEG) the mutation.

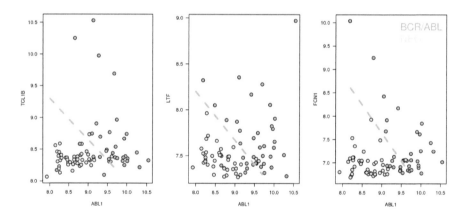

FIGURE 5.27: Some combinations of genes that were ranked high using LASSO. The most important gene ABL1 is combined with genes TCL1B, TLF and FCN1, respectively. The samples are colored by the mutation status of the BCR/ABL fusion gene, with (BCR/ABL) and without (NEG) the mutation.

quent multivariate classifier will search for. However, if the selection was not too stringent, this counter-intuitive combination might still provide interesting results. The reason is that very simple approaches are often better in extracting the small set of relevant information from high-dimensional complex data. Although in theory a very predictive gene signature does not need to contain genes that are differentially expressed, it is unlikely that in reality such a signature will be applied in the clinic. The primary reason is that a gene signature should reflect a very strong and robust difference between the classes it aims to predict. This generally implies some degree of differential expression of the genes in signatures.

Jeffery et al.[145] compared various univariate feature selection methods on 9 different real-life microarray datasets. Their main conclusions is that the performance of the different approaches depended on the sample size and on the variance structure of the dataset. In studies with 10 or more samples per group, methods based on fold change ranking performed weakly. In studies with fewer sample per group, or with a high pooled variance, these fold change methods however worked quite well. Overall, the moderated t-statistic[153] appeared to be (i) frequently among the best preforming methods, and (ii) the most robust method across all sample sizes. We therefore recommend its use for univariate feature selection.

Independent or in combination with the classifier. In independent feature selection methods, the gene selection is separated from the sample classification. In dependent feature selection, in contrast, the feature selection is embedded within the classification algorithm. Consequently, the feature selection works in interaction with the classifier.

Again, both have their own advantages. A clear separation between the feature selection and classification algorithm allows more flexibility to combine different methods and allows an assessment of the relative importance of the two steps. On the other hand, a feature selection as a part of the classification algorithm may optimize the discovery of a gene signature as the features are selected in an optimal manner for the classification. A typical example of independent feature selection is the combination of a univariate selection method with various classifiers, and a typical example of a dependent selection approach is the variable importance returned by random forest[182].

Optimal number of genes. An important issue with classification is how many genes one needs for optimal classification. To avoid this complex problem, researchers sometimes just used an arbitrary number of genes, like the 20 top-ranked genes in a *t*-test. Luckily, this inappropriate approach is getting rare in the recent classification analyses.

A more logical and common approach is to run many cross-validations using different numbers of genes. Comparing the misclassification error rates between the various gene set sizes allows to decide which number of genes

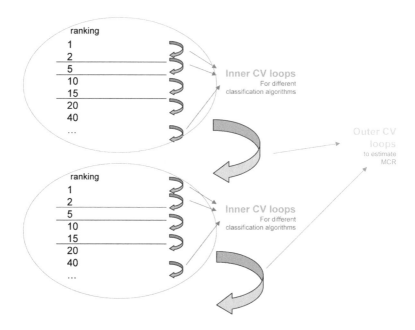

FIGURE 5.28: Scheme of nested loop cross-validation, showing that feature ranking is done only once in every outer cross-validation loop. Selection is done as many times as the user specified how many genes should be considered.

corresponds with the most accurate classification. Feature selection and classi-fication are typically combined to create a "complete classification procedure," and the outer cross-validations are performed with this complete classification procedure[184]. Consequently, feature ranking is done once in every outer cross-validation loop. Then, based on the cut-offs for x number of features pre-specified by the user, the top x features are used for inner cross-validation loops (see Figure 5.28). This is called "forward filtering."

Let's illustrate this concept by means of the BCR/ABL vs. NEG com-parison of the ALL data. Figure 5.29 shows the misclassification error in function of the number of genes used for classification. This has been done for PAM[185], random forest with variable importance selection[182] and NLCV (nested-loop cross-validation) with Limma to rank the genes (see Figure 5.28). A similar pattern can be seen in each of the three panels; an initial decrease in classification errors with increasing gene set size followed by an increase. As it is the misclassification error of the independent test samples that is being shown, this perfectly fits the theory of overfitting (see Section 5.6.2.1). It is exactly this dip that can be used to identify the optimal number of genes to use for classification. The three approaches select different numbers of genes,

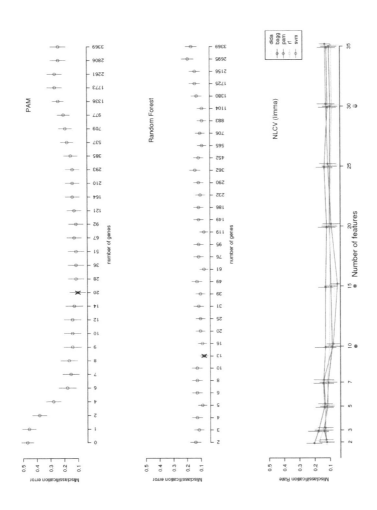

FIGURE 5.29: Optimal number of genes for classification. The misclassification error of the test samples is plotted in function of the number of selected genes using the training samples. This has been done for PAM (prediction analysis of microarrays), random forest with variable importance selection and NLCV (nested-loop cross-validation) using Limma to rank the genes.

which is not surprising given the strong conceptual differences between the approaches (see list at the end of Section 5.6.4.9).

5.6.4   Classification methods

There is a wide variety of classification algorithms that have been developed in very different research domains like statistics, data mining and machine learning. Some of the more popular methods for microarray data are listed below.

5.6.4.1   Discriminant analysis

5.6.4.1.1   Linear discriminant analysis   Linear discriminant analysis (LDA) is the oldest and one of the simplest classification methods[186]. LDA tries to find which linear combination of expression levels of the different genes best separates the two groups. In other words, it projects the high-dimensional data onto a line and performs classification in this one-dimensional space. The projection maximizes the distance between the means of the two classes while minimizing the variance within each class.

LDA assumes a multivariate normal distribution with equal group covariance, meaning that the distribution of gene expression measurements within the classes is normal and all classes have similar gene variances and gene-by-gene correlations. In terms of the different distance measures that can be used for estimate similarity, as discussed in Section 5.4.2.1.1, LDA makes use of the Mahalanobis distance.

There are many extensions on LDA (see [186],[187] for a review and more details). Diagonal linear discriminant analysis (DLDA) and diagonal quadratic discriminant analysis (DQDA) are simple discriminant rules for diagonal class covariance matrices with linear (DLDA) or quadratic (DQDA) discriminant function. PDA is a form of penalized LDA. It is designed for situations in which there are many highly correlated predictors[188]. Also, as with most linear models, LDA can be extended to non-linear models via the kernel trick (see StatsBox 5.22).

5.6.4.2   Nearest neighbor classifier

K nearest neighbor (k-NN)[189] is a non-parametric method using a measure of distance (see Section 5.4.2.1.1) between observations. The class of an unknown test sample is predicted by considering the k nearest training samples with known class. The most frequent class among the k neighbors is then the predicted class for that unknown sample. In other words, k-NN predicts unknown samples based on the assumption that the correct class is the most prevalent among other samples with similar gene expression profiles. It is therefore a straightforward algorithm with an appealing geometrical interpretation as it searches for local neighborhoods in gene expression profiles[190].

> **The kernel trick** is a method for using a linear model to solve a non-linear problem by mapping the original, non-linear, measurements into a higher-dimensional space, where the linear model can be used. This transformation allows to use a linear model (in the new space) which is equal to a non-linear model in the original space.

StatsBox 5.22: The kernel trick

The difficulty of kNN lies in the choice of k; how many neighbors should ideally be used? This can be determined by assessing the performance of different values of k (like 1, 2, 3, ..., N) on a training set by cross-validation (see StatsBox 5.4)[187].

5.6.4.3 Logistic regression

Logistic regression models the relationship between a binary response (a response taking two inputs, i.e., the two groups) and one or more explanatory variables (here the expression levels of a number of genes). It is a simple, straightforward model that is widely used and easy to extend. A logistic regression is a generalized linear model (GLM), which means that it is a generalization of the general linear model (see StatsBox 5.15 on page 174). This generalization is done by transforming the binary response to a normally distributed response with a logit function. This allows the use of the attractive, flexible modeling framework of general linear model. Logistic regression is also known as the softmax function in machine learning literature or as the multiple logit function[187].

Logistic regression models are typically fit by maximum likelihood, using (IRLS). Logistic discrimination provides a more direct way of estimating posterior probabilities and is easier to generalize than most other classification algorithms (see Section 5.6.5).

Logistic regression is known to perform bad when used on high-dimensional datasets[104]. The estimation of how each gene contributes to the class prediction can be quite inaccurate and unreproducible when there are more genes than samples. The "rule of 10" says that coefficients obtained with logistic regression may be biased if the number of events per explanatory variable falls below 10[191],[192]. To understand the impact on microarray data, imagine a classification problem where 50 genes are used to predict treatment response, where only 25% of the patients respond. This would require a population of 2,000 to obtain 500 responders (10 times the number of genes and 25% of the study population), which is generally not feasible[104].

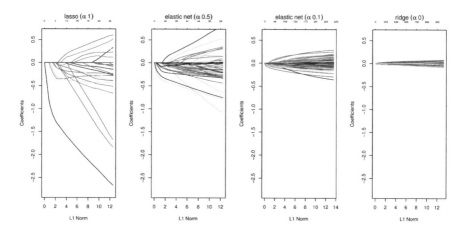

FIGURE 5.30: Penalized regression coefficients in relation to degree of penalization. From left to right, LASSO, elastic net (two alpha settings) and Ridge regression are shown.

Penalized logistic regression   To make logistic regression work on microarray data, it is advisable to include a penalization into the model, like Ridge (L_2 regularization[161]) or LASSO (L_1 penalization[185]) or elastic net (both L_1 and L_2 penalization[168]). Section 5.5.2.5.4 on penalized linear models describes each of these three methods, but the main points can be summarized as follows. Ridge regression (L_2 regularization) shrinks all coefficients to small non-zero values to prevent instability due to collinearity. LASSO (L_1 regularization) shrinks many, but not all, coefficients to zero while the other coefficients have comparatively little shrinkage. This can be used as a very interesting feature selection method; the genes with zero coefficients can be excluded because they are not informative. Elastic net, a combination of LASSO and ridge regression, gives intermediate results with fewer regression coefficients set to zero than in LASSO, but more shrinkage of the other coefficients[169].

This is illustrated in Figure 5.30 which shows the relation of coefficients with increasing penalization for the different algorithms using the ALL dataset. There is as expected a clear gradient from LASSO over elastic net to ridge regression. Initially (left in the graphs), all coefficients start at zero and the penalties are relaxed. With increasing penalties there are however two major differences between the methods. First, in ridge regression all coefficients become simultaneously non-zero, while in LASSO only a small number of coefficients become one-at-a-time nonzero. Second, LASSO contains a few genes with considerably high or low coefficients, while in ridge regression all genes have small non-zero coefficients.

TABLE 5.4: Number of selected genes and the top 7 genes for LASSO, elastic net (two alpha settings) and ridge regression.

Rank	LASSO		Elastic net (0.5)		Elastic net (0.1)		Ridge regression	
	Gene	Coeff.	Gene	Coeff.	Gene	Coeff.	Gene	Coeff.
1	ABL1	-2.66	TCL1B	-1.08	TCL1B	-0.36	ITGA5	-0.08
2	TCL1B	-1.84	RPS6	0.81	ABL1	-0.30	CCND2	-0.08
3	LTF	-1.67	ABL1	-0.78	ABL1	-0.29	TCL1B	-0.07
4	FCN1	-0.66	ABL1	-0.75	RPS6	0.29	FSTL1	0.07
5	DSTN	0.60	LTF	-0.65	FSTL1	0.25	ABL1	-0.07
6	RPS6	0.52	FSTL1	0.52	ABL1	-0.23	CGRRF1	-0.07
7	YES1	-0.51	ABL1	-0.44	LTF	-0.22	ABL1	-0.07
# selected	26		57		250		1008	

Table 5.4 shows the number of genes selected in each of the 4 penalized regression approaches. From the 3,369 genes used for the analysis, LASSO contains only 26 genes with non-zero coefficients, while ridge regression still has more than 1,000 genes. This illustrates that LASSO has very nice properties for gene selection. A closer look at the 7 genes with the highest absolute coefficients shows that the variation in coefficients between the genes also differs considerably between the methods. While ridge regression results in similarly large coefficients (around 0.07-0.08), there is a quite some variation between genes in LASSO. The gene called ABL1 with a coefficient of -2.7 towers above the others as the second-ranked gene only has a coefficient of -1.8. This can actually be seen as a kind of validation of the method, as the two groups being compared are ALL tumors with a different mutation in the BCR/ABL fusion gene. It is quite logical that a mutation in this gene primarily affects the expression of this particular gene.

Another solution is to work in a semi-supervised way (see StatsBox 5.3), where in a first unsupervised step the data are being reduced to a limited number of explanatory variables that can be used in a logistic model in the second step. Examples are supervised PCA[106] and gene harvesting[193].

5.6.4.4 Neural networks

A neural network (NN) or artificial neural network (ANN)[194] is a computer-based algorithm that is modeled in analogy with the structure and behavior of neurons in the human brain[195]. Basically, it tries to connect input (gene expression) and output (binary outcome) by building an intermediate layer (called hidden layer, see Figure 5.31). A key feature of ANNs is that they learn the relationship between inputs and output through training. Although this training can be unsupervised (see SOM in Section 5.4.2.2), it is most commonly supervised, as in classification problems.

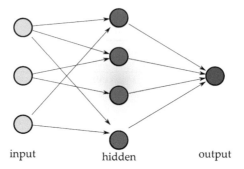

FIGURE 5.31: Scheme of a neural network. Neural networks build an intermediate, hidden, layer between input and output through training.

5.6.4.5   Support vector machines

Support vector machines (SVMs)[196] are used to detect and exploit complex patterns in data. An SVM constructs a hyperplane that separates the two groups in the high-dimensional microarray data space (see Figure 5.32). SVM avoids overfitting by choosing the widest plane from the many possible solutions (the so-called "maximum margin separating hyperplane," see Figure 5.32). The separating hyperplane can be stated entirely in terms of vectors in the input space (the so-called "support vectors") so that it can be located without the need to represent the entire microarray data space explicitly[197]. The kernel trick is widely used to apply linear classification to non-linear classification problems. Besides performing binary classifications, they can also be applied to multiclass and regression problems.

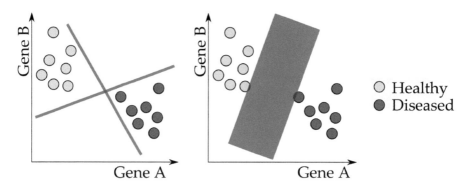

FIGURE 5.32: Schematic two-dimensional view of a support vector machine (SVM) model. SVM models a maximum-margin hyperplane and margins by supervised learning. Samples on the margin are called the "support vectors."

5.6.4.6 Classification trees

In classification tree structures, the leaves represent the classifications of the samples and the branches of the conjunctions of genes leading to those classifications. Tree construction usually comprises two steps: growing and pruning. The growing step begins with the so-called "root node," which is the entire learning dataset, i.e., all the samples used for training.

Generally, the tree is grown by partitioning the samples as far as possible. This in principle results in a tree with as many leaves as samples and with only one sample in each cluster. Such a saturated tree is generally too big and overfitted. This is why the saturated tree is pruned in a second step to obtain a reasonably sized tree that is still discriminating the two groups.

5.6.4.7 Ensemble methods

Ensemble methods are actually no classification models on themselves, but techniques to improve the accuracy of a classification model by combining many individual models in an "ensemble." More specifically, they combine the predictions of multiple, individually trained classification models to predict independent samples. They can be used in principle with any classification technique.

5.6.4.7.1 Bagging Bagging (bootstrap aggregating)[198] produces replications of the training set by sampling with replacement, which is called "bootstrapping" (see StatsBox 5.23). Each replication of the training set has the same size as the original at all. A classifier is generated from each replication, and all classifiers are used to classify each sample from the test set using a voting scheme.

Bagging should only be used if the learning machine is unstable, like neural networks or classification trees. A learning machine is said to be unstable if a small change in the training set yields large variations in the classification.

5.6.4.7.2 Boosting Boosting[200] is a family of methods for accelerating a learning algorithm. The idea is to boost a weak learning algorithm into a strong learning algorithm. A weak learning algorithm is a classifier which is only slightly correlated with the true classification. For example, it can be weak due to inaccurate rules of thumb which are only slightly better than random guessing.

Most boosting algorithms are an iterative production of many classification models. Each classification depends on the previous one, where more weight is given to the samples that were misclassified. More specifically, the training samples chosen at a given time depend on how well they performed in earlier classifications. Samples that were harder to classify will be used more frequently, so that the algorithm is concentrating on the more difficult samples.

Jackknife and bootstrap. Bootstrap and jackknife are re-sampling techniques used to assess the errors in a statistical estimation problem. As they are nonparametric, they have the advantage that they are easy to describe and that they can be applied to complex datasets without having to make distributional assumptions like normality[199].

Jackknife consists out of multiple loops, say n loops. In each loop, one observation is left out from the sample set, creating n data subsets. Afterwards, the statistic of interest (like an average) is calculated on each subset. The variance of these "re-sampled averages" is then an estimate of the variance of the original sample mean.

Bootstrap, which is basically a generalization of the jackknife, estimates the sampling distribution of a statistic by sampling data subsets with replacement from the original sample[a].

---

[a]The words jackknife and bootstrap are both metaphors for tools to help oneself without external help when in trouble. A jackknife is a pocket knife, and a bootstrap is the loop on the top of tall boots that allows to easily pull on the boots. Indeed, these re-sampling methods can be of great help as they allow to estimate parameters rather easily without the need to make distributional assumptions.

StatsBox 5.23: Jackknife and bootstrap

5.6.4.7.3 Random forest

Random forests[201] is a very different ensemble method of classification trees compared to bagging or boosting.

Each of the classification trees is built using a bootstrap sample of the data, and at each split the candidate set of variables is a random subset of the variables. Thus, random forest uses both bagging (bootstrap aggregation), a successful approach for combining unstable learners[164], and random variable selection for tree building.

Each tree is unpruned (grown fully), so as to obtain low-bias trees; at the same time, bagging and random variable selection result in low correlation of the individual trees. The algorithm yields an ensemble that can achieve both low bias and low variance because it is an average over a large ensemble of low-bias, high-variance but low correlation trees.

Random forest has excellent performance in classification tasks, comparable to support vector machines[182]. From the perspective of prediction, random forests is about as good as boosting, and often better than bagging[202].

5.6.4.8 PAM

Prediction analysis for microarrays (PAM)[185] is a classification technique using "nearest shrunken centroids." Basically, PAM gradually shrinks the gene centroids (i.e., the average gene expression levels) of the two groups towards the overall centroid. The distance between the centroids of two groups will be small when the respective gene is not differentially expressed. Their group centroids will therefore quickly reach the overall centroid so that these genes are removed first. Differentially expressed genes, in contrast, will survive the shrinkage much longer and may end up in the final selection of genes used for classification. This ranking of genes by PAM can be seen as a type of penalized t-statistic (see Section 5.5.2.2).

The optimal amount of shrinkage can be found with cross-validation, and this is used to select the number of genes to use for classification. At the end, PAM uses the centroids of these genes to classify new samples to the nearest centroid.

5.6.4.9 Comparison of the methods

There is a wide variety of classification algorithms, and many of them have profound conceptual and methodological differences. It is therefore not an easy task to make general statements about which classification method performs best in which situation. Given the differences between the methods, it can be expected that no single classification method works is always the best, but that certain methods perform better depending on the characteristics of the data.

In the ideal case with a strong signal, most or all classification methods are however expected to provide similar classification accuracies, perhaps with using slightly different genes. The reason is that, irrespective of differences in underlying mechanisms, these methods are all designed for the same purpose, namely maximizing classification accuracy. They should consequently pick up a very strong signal when present. This implies that a discovery of a signature with only one single classification method is worrying, and should be carefully examined.

Personally, we like to apply four different approaches:

1. PAM[185] applying univariate and dependent feature selection.

2. Random forest with variable importance filtering[201],[182] applying multivariate and dependent feature selection.

3. LASSO[185] or elastic net[168] applying multivariate and dependent feature selection.

4. Forward filtering in combination with various classifiers (like DLDA, SVM, random forest, etc.) applying an independent feature selection. The selection can be either univariate or multivariate depending on the

chosen selection algorithm; we usually choose Limma as a univariate and random forest variable importance as a multivariate method.

All four methods have the property that they search for the smallest set of genes while having the highest classification accuracy. The underlying rationale and algorithm is very different between the four approaches, making their combined use potentially complementary. All these methods have been illustrated using the ALL dataset (the classification of BCR/ABL and NEG samples) in the previous sections.

5.6.5 Complex prediction problems

5.6.5.1 Multiclass problems

Multiclass classification can be constructed often as a generalization of binary classification. Examples are multiclass logistic regression, multiclass SVM and multiclass random forest. In practice, however, multiclass problems are often reduced to several two-class problems. One either analyses all pairwise comparisons, or one compares one group vs. the pool of all other classes (one-vs.-all). Ramaswamy et al.[203] combined support vector machines (SVMs), which are binary classifiers, to solve the multiclass classification problem. They showed that the one-vs.-all approach of combining SVM yields the minimum number of classification errors on their Affymetrix data with 14 tumor types.

5.6.5.2 Prediction of survival

Clinical data often are in the form of time-to-event data. Examples are time to death in fatal diseases, or time to relapse after therapy of diseases that are difficult to cure like cancer or depression.

Ideally, an analysis would use all the information of the time-to-event data by using survival analysis approaches (like Cox proportional hazards models [204]). In practice, however, one generally creates two (or more) classes by cutting the survival times at the median survival time or another cut-off value [106]. Subjects living longer or less than the cut-off are then categorized in two different groups. Here, these two groups could for example be named the low and the high mortality group, respectively.

5.6.6 Sample sizes

The fact that there are more variables (genes) than measurements (samples) lies at the heart of most issues with classification. It is often forgotten that multivariate techniques originally crashed when there were more variables than measurements. Quite logical, as not all variable combinations could be evaluated due to a too small sample size. Although sample sizes in microarray classification studies clearly continue to increase because of the lower costs and the increased acceptance of its relevance, it is unlikely that single studies will contain more samples than genes as even large studies as phase III clinical trials rarely exceed 5,000 subjects. Classification exercises will therefore always suffer from small sample sizes. Where sample size calculation for tests of differential expression was already complex and dependent on multiple assumptions, sample size calculation for classification are even more cumbersome. A common pragmatic approach is therefore to let the budget decide with the attitude "the more the better." As discussed in Section 5.6.2.3, a larger sample size will lead to a more accurate prediction and to a more realistic estimation of the prediction accuracy. Because classification inference is situated more at the individual level rather than at the population level, it is of course easy to add extra samples at later stages to the test data.

5.7 Pathway analysis

The detection of differentially expressed genes (see Section 5.5 on page 153) suffers from multiple problems due to the high-dimensionality of microarray data and due to the fact that a gene level focus does not show "the bigger picture" on the patterns in the data. Subramanian et al.[205] listed these limitations as follows:

1. Genes that are actually biologically relevant may turn out to be nonsignificant, primarily due to small sample sizes and multiplicity corrections.

2. Long lists of significant genes are difficult to interpret without any unifying biological theme.

3. Analyzing genes separately ignores the reality, in which genes do not work in isolation but in pathways.

4. Top lists of significant genes can be quite unstable. Although these genes may have similar significance levels in replicated experiments, small differences in their ranks may result in disturbingly small overlaps between reproduced gene lists. Analogously, subtle deviations in the analysis strategy, like the use of a different summarization technique, regularly result in different gene lists[206].

Moving the level of analysis from single genes to sets of related genes overcomes many of these statistical challenges. Such "gene sets" are defined based on prior biological knowledge, like published information about biochemical pathways or co-expression in previous experiments[205]. Moreover, such a pathway-oriented approach allows to include previously accumulated biological knowledge into the analysis making the analysis more biology-driven[112].

Logically, the main weakness of this approach is its dependence on the quality of the knowledge used to define the gene sets. Pathway analysis tools rely heavily on the existing functional annotation, which can be assumed incomplete for most organisms of interest[87]. However, despite its inextricable incompleteness, the gene set definitions may be relevant or comprehensive enough to guide researchers more easily to interesting but concealed patterns in the data.

5.7.1 Statistical approaches in pathway analysis

Pathway analysis of gene expression data is a complex and relatively young research field with a high impact on drawn inferences. This fueled the development of many analysis methods with widely different reasonings and assumptions. There are basically three main strategies to analyse pathways: over-representation analysis, functional class scoring or gene set analysis.

5.7.1.1 Over-representation analysis

Over-representation analysis (ORA) searches for gene sets that are enriched with significant genes. It starts from a list of (differentially expressed) genes and tests whether the gene set is over-represented in this list. More specifically, it calculates for each gene set the probability that it contains more differentially expressed genes than would be expected by chance.

Imagine for example a situation where 1,000 genes were tested for differential expression, and where 100 genes have been selected as significant. A hypergeometric test would indicate that a gene set of 10 genes has a probability of 0.01 to contain 3 significant genes and only a probability of 0.0001 to contain 5 significant genes. The smaller the probability, the more evidence there is that the concerning gene set is truly enriched with differentially expressed genes.

There are many ORA approaches (see [207],[208] for a list and review). They are either based on parametric test (χ^2-test and the binomial test z-test) or on non-parametric alternatives (hypergeom[208]). All these methods are identical when the sample sizes are sufficiently large[207],[112],[208]. In this scenario, parametric tests might be preferred because their calculations are faster. Non-parametric approaches are computationally more intensive, but are recommended for small sample-sized datasets, where the assumptions underlying parametric tests are usually not satisfied. Since microarray experiments generally involve small datasets, and given the currently available computing means, the non-parametric hypergeometric test is therefore generally used as the standard approach[208].

Often an ORA analysis using GO definitions (see Section 5.7.2) identifies multiple, directly related gene sets with considerable overlap of genes. This is because of the hierarchical nature of GO gene sets where each "parent" term inherits all annotations from each of its "children"[209]. To alleviate this problem, Falcon and Gentleman[210] have implemented a conditional hypergeometric test that uses the relationships among GO terms to decorrelate the results.

5.7.1.2 Functional class scoring

Functional class scoring (FCS) searches for gene sets with relatively low average p-values[211],[212]. It starts from a comprehensive gene list of p-values and calculates for each predefined gene set a score that summarizes the significance of all the genes included. Typically, the mean of the negative log-transformed p-values is used to summarize the significance score of a gene set[211],[212]. Next, it is calculated how likely it is to observe such a *-log(p)*. In practice, this is done by converting the summarized score for each gene set into a p-value. Very small p-values indicate that it is unlikely to observe such a *-log(p)* just by chance, suggesting that the gene set is truly enriched with differentially expressed genes. This conversion is done by comparing the obtained scores to an empirical null distribution reflecting *-log(p)* values that one expects to see just by chance. The empirical null distribution is derived by either bootstrapping randomly selected gene sets of the same size[211] or by permuting the samples[212].

Both FCS and ORA use the same input, namely a list of ranked genes with p-values. However, ORA methods cut the list into significant and non-significant genes, and use only the former for the pathway analysis. As FCS does not select genes a priori, it allows to incorporate all available information from the gene list. First, it uses not only a subset of genes but all the genes measured with the microarray. Second, it does not categorize genes into significant or not, but incorporates the quantitative nature of the p-values. This latter is not only more informative but also more representative. Imagine three genes: two significant ones with p-values of 0.049 and 0.0001 and a non-significant one with a p-value of 0.051. While an ORA method will bin the two significant genes together, a FCS method will correctly regard the genes with p-values of 0.049 and 0.051 as quite similar, and the gene with 0.0001 as relatively much more significant. This example highlights the main drawback of ORA; its results may depend too strongly on the threshold used to select the genes. As expected, FCS therefore yields more consistent results than ORA[211].

5.7.1.3 Gene set analysis

Gene set analysis (GSA) searches for differentially expressed gene sets. GSA directly scores pre-defined gene sets for differential expression and especially aims to identify gene sets with "subtle but coordinated" expression changes[213].

GSA methods are based on widely different methodological assumptions[112]. Based on how the null hypothesis is defined, GSA methods can be categorized into competitive, self-contained and mixed methods.

5.7.1.3.1   Competitive methods   The competitive methods test the relative enrichment of differentially expressed genes in a gene set compared with a standard, which can be defined as a background set, but mostly as the pool of all other gene sets. Its null hypothesis is consequently

H_0^{comp} : *The genes in G are at most as often differentially expressed as the genes in G^c*

With G being the gene set of interest and G^c the pool of all other available gene sets. The competitive behavior can introduce the so-called "zero-sum" behavior because gene sets are only evaluated in reference to other gene sets in the data and not to an independent reference[214]. For example, in an extreme experiment in which almost all genes are down-regulated, some gene sets may be considered up-regulated[213].

Competitive methods generally randomize genes across gene sets to derive the standard for determining the relative enrichment of differentially expressed genes[112]. This gene-sampling models implicitly assume that the genes are independent, which is evidently false. Goeman and Bühlmann[112] have shown that gene-sampling may call gene sets falsely significant when containing correlated genes. Some gene sets containing correlated genes but without a differentially expressed one can for example be called significant. As gene sets typically contain genes with correlated expression profiles, Goeman and Bühlmann[112] therefore correctly recommend against the use of gene-sampling models.

See the review of Nam and Kim[213] for a list of available competitive methods.

5.7.1.3.2   Self-contained methods   Self-contained methods use only the information contained in a given gene set, and compare this gene set to a fixed standard that does not depend on the measurements of genes outside the gene set[112],[213].

H_0^{self} : *No genes in G are differentially expressed*

This null hypothesis that no single gene in a set is differentially expressed is a strong statement. Consequently, gene sets will easily be detected as being enriched with differentially expressed genes. Sometimes the method might be even too powerful. As only a single differentially expressed gene can make the whole gene set significant, a gene set may be significant while not really being "enriched" with differentially expressed genes.

The most frequently applied self-contained tests are the global test[215], global ancova[216] and PLAGE[217].

5.7.1.3.3 Hybrid methods Hybrid methods do not test individual gene sets, but the entire population of gene sets. They use a competitive statistic to score the relative enrichment of the gene sets, and subsequently test the significance of the entire dataset by applying sample permutation to the scores. Hybrid methods are considered a competitive method relative to individual gene sets, but considered a self-contained method relative to the entire dataset (set of gene sets)[213].

The most popular hybrid method is GSEA[205]. Efron and Tibshirani[218] introduced GSA, an improved version of GSEA, by adding a more generally sensitive statistic to score the gene sets, the maxmean. There are however still some concerns with the use of GSEA; Damian and Gorfine[219] showed that the enrichment score provided by the GSEA procedure can be influenced by the size of a gene set and by the presence or absence of lower-ranking sets.

5.7.1.4 Comparison of the methods

Primarily due to the complexity and the temporariness of pathway data bases, there is no standard statistical approach in place yet to analyse pathways. A consensus of opinion is however arising that some methods like hypergeometric tests are suboptimal while others like gene set analysis seem very promising[214],[213].

ORA methods are suboptimal as they are based on an arbitrary cut-off based selection of genes. Pavlidis et al.[211] emphasize the advantage of considering all available genomic information rather than sets of genes that pass a predetermined threshold of significance. GSA and FCS methods are free from the problems of the "cutoff-based" methods[213].

An important problem with both ORA and FCS, and some of the GSA methods, is that the statistical methods applied are based on the wrong assumption of independent gene (or gene group) sampling, which increases false positive predictions[112],[213].

Example. Let's use the comparison between the T-cell samples and the B-cell samples of the ALL dataset. Pathway analysis with MLP gives much more interpretable information compared to gene lists (see Table 5.5). Figure 5.33 shows the same results, but visualized in a pathway to make the interconnections between the different gene sets more clear.

Previously, based on a spectral map (Figure 5.11), we already briefly explored some of the key genes driving the difference between the T- and B-cell samples. We mentioned the genes TCF7 and CD3, known to be specifically expressed by T-cells[142],[141], which were indeed up-regulated in T-cells and genes over-expressed in B-cells (like HLA-DMB, HLA-DRB5, etc.) that apparently all belonged to histocompatibility class II molecules. This rather elaborate recapituation of these previously obtained results makes clear that the process to identify interesting candidate genes and the process to put them in a biological context is rather cumbersome and difficult.

TABLE 5.5: The top gene sets selected by MLP.

GO class	GO term	p-value
GO:0019882	antigen processing and presentation	1.71E-14
GO:0046649	lymphocyte activation	6.93E-12
GO:0045321	leukocyte activation	1.32E-11
GO:0001775	cell activation	3.32E-11
GO:0002504	presentation via MHC class II	8.23E-11
GO:0006955	immune response	9.35E-09
GO:0030098	lymphocyte differentiation	5.04E-08
GO:0042110	T cell activation	7.40E-08
GO:0045058	T cell selection	1.44E-07
GO:0002521	leukocyte differentiation	2.96E-06
GO:0030217	T cell differentiation	3.40E-06

With MLP, we receive directly a high-level picture of which biological pathways seem to be most affected. Interestingly, these gene sets seem to be biologically quite relevant as they indicate processes as T-cell activation and selection and other processes related to immune response and lymphocytes.

Pathway analysis with GSA gives highly similar output as with MLP (see Table 5.6). Almost the same gene sets, including T-cell activation and selection and processes related to immune response and lymphocytes, are topranked. While MLP only looks at significance regarding which direction, GSA incorporates the direction into the analysis. This is why GSA discovers the B-cell related processes, which are in the opposite direction but not as significant. Which approach is best is not yet clear to us. GSA has the advantage that it can discover more subtle effects in the other direction than where the most effects are heading to. In this situation this nicely resulted in the identification of gene sets specific for B-cell samples. On the other hand, Figure 5.34 demonstrates that a gene set (such as `lymphocyte activation`) can be significant in one direction (up-regulated in T-cells) while containing genes that are significant in the other direction (i.e., the genes from the child gene set `B cell activation` and `lymphocyte proliferation` that are down-regulated in T-cells). This indicates that ignoring direction makes sense in pathway analysis as gene sets can contain interesting patterns in both ways.

Figure 5.34 shows the same results, but the gene sets are now visualized in a pathway so that their interconnections can be studied more closely.

5.7.2 Databases

Besides the gene expression measurements from a microarray experiment, the second necessary input for the different statistical pathway analysis approaches discussed above is the definition of gene sets. To be able to check whether a certain pathway or set of genes is more affected in an experiment,

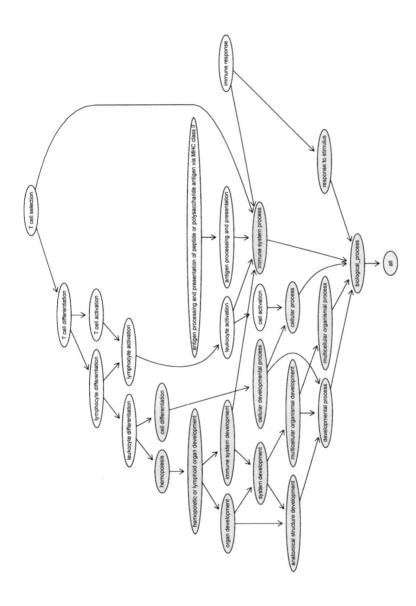

FIGURE 5.33: GO pathways containing the 11 top-ranked gene sets with MLP. The 11 most significant gene sets are highlighted in green.

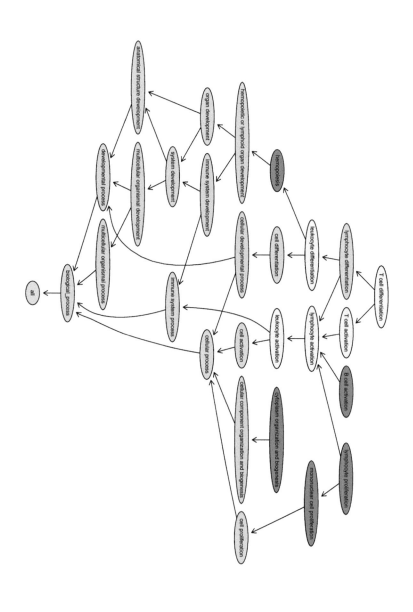

FIGURE 5.34: GO pathways containing top-ranked up- and down-regulated gene sets by GSA. The 5 gene sets most upregulated in T cells are highlighted in green, the gene sets most upregulated in B cells are highlighted in red.

TABLE 5.6: Top 5 up-regulated and top 5 down-regulated gene sets selected by GSA.

	GO class	GO term	Score	p-value
up-regulated				
in T cells				
	GO:0030217	T-cell differentiation	0.99	0
	GO:0042110	T-cell activation	0.85	0
	GO:0045321	leukocyte activation	0.49	0
	GO:0046649	lymphocyte activation	0.51	0
	GO:0002521	leukocyte differention	0.43	0.0004
up-regulated				
in B cells				
	GO:0019882	antigen processing and presentation	-2.50	0
	GO:0042113	B-cell activation	-1.08	0
	GO:0032943	mononuclear cell proliferation	-0.63	0.03
	GO:0046651	lymphocyte proliferation	-0.63	0.03
	GO:0007028	cytoplasm organization	-0.98	0.03

one needs to have access to databases that provide such definitions of gene sets.

Most statistical pathway analysis currently makes use of gene set definitions from a single source. Often the databases are updated regularly so that the content quality reflects current knowledge. To avoid issues with, e.g., outdated information, it is adviseable to retrieve the gene set information directly from the source. For example definitions of gene sets about biological processes as defined by the gene ontology (GO) consortium should be obtained from GO directly and not, e.g., via Entrez gene.

It is understandable that single databases such as the ones mentioned below do not reflect the complete biological knowledge. Therefore, pathway analysis should benefit from an integration of the different databases. However, a database that would integrate the information from multiple sources can be difficult to design and requires a substantial effort in keeping the content current as changes in the various data sources demand regular updates of the integrating database. Due to these difficulties, most scientists still make use of single gene set definitions. Most common is the use of the definition of "biological process" gene sets from the GO consortium.

Below are a number of gene set databases mentioned. This list is certainly not comprehensive, but gives a flavor of the types of databases that can be used.

The different binnings (such as GO tree BP MF CL, KEGG biochemical pathways, chromosomal location) are suitable depending on the experimental context.

5.7.2.1   Gene ontology

In this large international initiative, researchers are looking into describing genes and their gene products according to three major themes using standardized phrases ("controlled vocabulary"). For each gene product they aim to assign one or many characteristics based on the *biological processes* they are involved in (e.g., growth, reproduction, metabolism, etc.), what *molecular function* they have (e.g., catalytic activity, transporter activity, binding, etc.) and where they are located within the *cellular components* (e.g., synapse, extracellular matrix, envelope, etc.). Using such standardized vocabulary leads to the creation of groups of gene products that share the same characteristic. This grouping can in turn be utilized in gene set or pathway analysis algorithms. The ontologies are continuously updated, and new versions are made available on a monthly basis through the GO website.

5.7.2.2   KEGG

The *Kyoto Encyclopedia of Genes and Genomes*[220] was initiated by the Japanese human genome programme in 1995 and comprises a set of online available reference databases that deal with genomes, enzymatic pathways and biological chemicals: KEGG PATHWAY is the major component of KEGG. It contains a collection of manually drawn biochemical pathway maps showing molecular interaction and reaction networks. In total, six different areas are covered: metabolism, genetic information processing, environmental information processing, cellular processes, human diseases and drug structure relationships. KEGG GENES contains data on genetic building blocks of genes and proteins. KEGG LIGAND focuses on chemical building blocks of both endogenous and exogenous substances. KEGG BRITE covers hierarchies and relationships of various biological objects.

The different KEGG databases can be used to link genomes to biological systems and environments. Such tools are essential for the analysis of gene functions in the context of the networks of genes and molecules. The KEGG databases are daily updated and made available free of charge[221].

5.7.2.3   GenMAPP

The gene map annotator and pathway profiler (GenMAPP) is a Windows-compatible computer software for analyzing gene expression data. The analysis takes scientific knowledge about diseases, subcellular localization, biological functions and pathways (e.g., signal transduction pathways, metabolic pathways, etc.) into consideration by grouping genes according to this information. GenMAPP is an open-source bioinformatics application that was originally developed in the group of Bruce Conklin at the J. David Gladstone Institutes in San Francisco in collaboration with Cytoscape in 2000. It nowadays contains information on 11 species to support the visualization (e.g., covisualization of pathways and gene expression data from microarrays) and analysis of genomic data. The project still maintains a number of databases (gene identifiers, pathway maps, etc.) as well as the different visualization and analysis tools (including tools for drawing pathways from scratch based on the scientific knowledge of the user).

5.7.2.4 ARED

A total of 5 to 8% of all human genes contain a 13 bp adenylate-rich element (ARE) in the 3'UTR. This element is important in the mediation of the turnover of mRNAs that regulate among others cellular growth, differentiation and the response to external stimuli (e.g., microbes, inflammatory response, environmental alterations, stress agents, etc.). The adenylate-rich element database (ARED) contains mRNAs with the ARE sequence. Genes like interferons, cytokines and proto-oncogenes are known to contain the ARE motif.

5.7.2.5 cMAP

Concept maps are another way of capturing information about gene networks that can be utilized for obtaining gene sets. In essence, tools such as CmapTools (developed at the Institute for Human and Machine Cognition – IHMC) provide a software environment that allows a user or a group of people to collaboratively construct, share and publish knowledge about a specific scientific domain. This is made possible by tools that aid in the construction and modification of the maps. Furthermore, there are public concept map servers (CmapServers) that have been created to promote the sharing of knowledge in different domains. These servers also offer the possibility to link concept maps to other related concept maps.

5.7.2.6 BioCarta

BioCarta was founded in April 2000 by a team of scientists and engineers. Via the Internet site of BioCarta, scientists can obtain and share information on gene functions, proteomic pathways (Figure 5.35 shows the cdc25 and chk1 regulatory pathway as an example of how these pathways are visualized), and exchange research reagents. The site is moderated by scientists and does not ask for any usage fees.

5.7.2.7 Chromosomal location

There are some experimental conditions (e.g., during embryonal development) when genes in distinct chromosomal regions are affected. To be able to detect modest but coordinated/co-regulated changes in expression of genes that share a common chromosomal location, one can define gene sets based on chromosomal location.

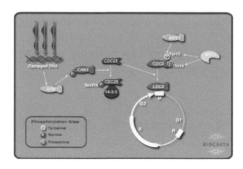

FIGURE 5.35: Example of a BioCarta pathway. Visualized above is the cdc25 and chk1 regulatory pathway in response to DNA damage. Image courtesy of BioCarta.

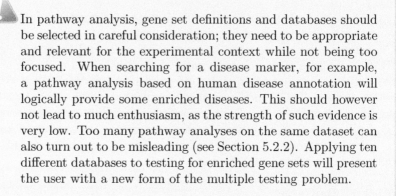

In pathway analysis, gene set definitions and databases should be selected in careful consideration; they need to be appropriate and relevant for the experimental context while not being too focused. When searching for a disease marker, for example, a pathway analysis based on human disease annotation will logically provide some enriched diseases. This should however not lead to much enthusiasm, as the strength of such evidence is very low. Too many pathway analyses on the same dataset can also turn out to be misleading (see Section 5.2.2). Applying ten different databases to testing for enriched gene sets will present the user with a new form of the multiple testing problem.

5.8   Other analysis approaches

5.8.1   Gene network analysis

Large-scale gene expression measurements raise the possibility of extracting information on the underlying genetic regulatory interactions directly from the data. This search is very valuable as it may help to unravel protein-to-protein, protein-to-DNA, or co-expression patterns. There is however no such thing as a free lunch. Evidence for such detailed information needs to come from large sample-sized studies. Moreover, given the curse of high-dimensionality (see Section 5.2) and the complexity of the aimed patterns, many interactions will be discovered with microarray data but a distressingly high proportion of these discoveries will be false positives.

A gene interaction network is a graph in which the nodes correspond to genes, and with edges between two nodes if the genes interact. A group of related genes corresponds to a connected subgraph of the interaction graph. Ideker et al.[222] assign to each subgraph a score (typically the sum of the scores of its component genes) and use a search algorithm to find subgraphs with high scores.

Nacu et al.[223] provide a framework to construct an interaction network based on single-gene scores like obtained from a test for differential expression. It allows to combine the interaction network approach[222] with pathway analysis (GSEA[205]).

More commonly, genetic regulatory networks are searched directly from the large-scale expression data using unsupervised techniques like spectral maps (see Section 5.4.3) or biclustering (see Section 5.4.2.2.6). Such techniques are indeed designed to find genes with correlated expression, and one hopes that the correlation may indicate co-regulation. Another potential statistical approach towards gene network inference is multiple regression (see Section 5.5.2.5), by which the regulation of each gene is modeled by multiple regulatory inputs. Clearly this is a very ambitious goal and to be successful it will correspondingly need very large sample sizes and/or a small number of predefined genes[224].

Groups of interacting genes are often identified based on observed correlations in gene expression between the genes. If the expression of gene A always increases when gene B decreases, they might be part of the same pathway or that they may regulate each other. However, such searches will always contain a considerable amount of false positives.

Furthermore, correlation does not imply causality. A correlation in gene expression between two genes does not imply that one causes the other. Let's illustrate this with an example in a totally different context. If the number of people buying ice cream at the beach is statistically related to the number of people who drown at the beach, then nobody would claim ice cream causes drowning because it's obvious that it isn't so. Both drowning and ice cream buying are clearly related by a third factor, i.e., the number of people at the beach. If an unknown factor or gene is affecting gene A and B, the correlation of A and B does not indicate a direct interaction between them.

5.8.2 Meta-analysis

With the increasing availability of cancer microarray datasets there is a growing need for integrative computational methods that evaluate multiple independent microarray datasets investigating a common theme or disorder. Meta-analysis techniques are designed to overcome the low sample size typical to microarray experiments and yield more valid and informative results than each experiment separately[225]. A meta-analysis is defined as a statistical approach that combines the results of several studies that address a set of related research hypotheses.

A meta-analysis of two cancer microarray datasets[226],[204] identified a robust gene signature for the prediction of normal and cancer samples[225]. The two studies used different Affymetrix platforms, namely the Hu6800[204] and the HG_U95Av2[226] chip types. Other examples of meta-analysis studies on cancer studies are Chan et al.[227] and Wirapati et al.[228].

If all the different studies would have been using the same platform, we would start from the raw data and normalize all samples together to obtain a more harmonized dataset. We would subsequently analyse all the data together in a linear model, while introducing study as a covariate in the model. The latter would allow us to test for the existence of significant between-study variation and simultaneously to correct for it (if present). Meta-analyses on the summary statistics of the different studies is not recommended since in microarray research the individual data are often available. Many microarray studies make their raw data publicly available. Hence, analyzing all the raw data simultaneously is preferred as it introduces more information into the analysis.

A typical potential weakness of a meta-analysis is its dependency on published studies. As it is more difficult to publish studies that show no significant results, the meta-analysis may be based on a biased sample of all studies conducted in the context. This publication bias or "file-drawer effect" (where non-significant studies end up in the desk drawer instead of in the public domain) should be seriously considered when interpreting the outcomes of a meta-analysis.

5.8.3   Chromosomal location

While gene expression analysis is currently focused primarily on the identification of differentially expressed genes, only few papers have looked into links between differential gene expression and the chromosomal localization of the genes (see Figure 5.36). Techniques such as MACAT are emerging which identify differentially expressed chromosome regions[229]. Especially experiments using cancer samples should benefit from such analyses as chromosomal instabilities such as deletions or amplifications of whole regions are regularly encountered in malignant tumors[230]. These copy number changes should be reflected in the measured gene expression intensities. Another application for this analysis approach is the search for groups of genes whose transcriptional activity may be influenced by their chromosomal localization[231].

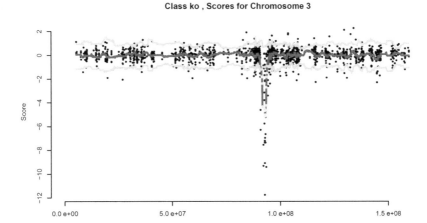

FIGURE 5.36: Identification of differentially expressed chromosomal regions. Yellow dots mark genes with statistically significant differential expression.

Chapter 6

Presentation of results

This chapter provides an overview of the typical graphs and visualizations used in microarray analysis. The outlook of a graph primarily depends on the information and message it is aimed to visualize. There are working figures to help the data analysts themselves understand the data better, and there are presentation graphs to present certain data characteristics to others. Plots generated for data analysis purposes may be rich in detail and complex to understand, as long as they facilitate the interpretation of the data. Plots for presentations to a larger audience, in contrast, should be very limited in content and should ideally convey only one message at a time.

For example, a heatmap presenting all genes and all samples can indicate which genes contribute to which sample clusters, and will show how similar or dissimilar the samples and the genes are to one another. Such a graph is typically intended for own use to explore the data. The same heatmap could also be used in a presentation, but should then be focused on only one aspect. If the purpose is to highlight that a certain selection of genes can discriminate two groups, it is advisable to include only these selected genes in the heatmap. If the purpose is to highlight the overall similarity of samples within certain groups, it would be wise to remove the individual gene names and to color the samples by their respective group. If interesting, some of the genes driving this grouping may be highlighted in a separate graph that shows the expression profiles of these genes across the relevant samples.

6.1 Data visualization

6.1.1 Heatmap

A heatmap is a two-dimensional false color plot visualizing the expression levels of many genes across several samples. Every row represents a gene expression profile across the samples (the columns). The abundance is depicted by color-coded squares. In the microarray world, the plot has become well known as Eisen plot[113]. Originally the color coding went from green over black to red, whereby green represented down-regulated genes, black unchanging genes and red depicted up-regulated genes. As this type of color-coding is not always easy to look at, many people have switched to a color coding that is more intuitive: red (intuitively representing hot or high) refers to up-regulated genes, white depicts genes that do not change, and blue (intuitively representing cold or low) depicts down-regulated genes (see Figure 6.1). This is because red and green are difficult to distinguish by color-blind people (which constitute 5% of the male population).

A heatmap typically has a dendrogram at the side and/or the top to indicate the relationship between, respectively, the genes or the samples. The ordering of the genes/samples is then based on the clustering of the dendrogram, so that similar genes/samples are close to each other. As a result, the heatmap will have a more homogeneous color pattern which facilitates its interpretation. Sometimes, however, one prefers to impose a certain ordering of genes or samples. When a heatmap shows for instance a top list of genes, the researcher may want to sort the genes from most to least significant. Or in a time series experiment, a researcher may want to sort the samples according to time.

Example. Let's interpret the heatmap shown in Figure 6.1. This heatmap visualizes the expression levels of 100 selected genes for all samples from the GLUCO dataset. The selection is based on differential expression between the 6 samples receiving a glucosamine treatment (Gln +) and the 7 samples that did not receive glucosamine (Gln -) using a moderated t-test (Limma[156], see Section 5.5.2.2.3). More specifically, Figure 6.1 shows the 100 most differentially expressed genes. What catches the eye is the clear clustering of the control samples (IL1-β - and Gln -), colored in red on top of the heatmap. The three other treatment groups have more similar gene profiles. This may suggest that it is mainly the control group that is driving the selection when testing for glucosamine effects. The 100 genes nicely separate the samples treated with (Gln +) or without (Gln -), which is as expected as these genes were selected to this end. More surprising is that this gene set also discriminates samples receiving a IL1-β treatment (IL1-β +) from those not receiving (IL1-β -).

The gene clustering at the right of the heatmap indicates the presence of four groups of genes with distinct expression profiles across samples (groups A to D). Compared to the control, the genes of group A and B are down-regulated, while the genes of group C and D are up-regulated in the three treatments compared to the control. The genes from group B have similar expression levels in all treatments except the control, while the genes of group A have lower expression levels in the combined glucosamine/ IL1-β treatment (IL1-β + and Gln +). Group D differs from group C due to lower expression in the samples treated with IL1-β treatment but without glucosamine (IL1-β + and Gln -). These four distinct profiles are illustrated in Figure 6.2 by four intensity plots that show the expression of a representative gene for each of the four groups.

As discussed in Section 5.4.2, the dendrogram will cluster samples/genes differently when using different linkage or distance measures. We suggested in

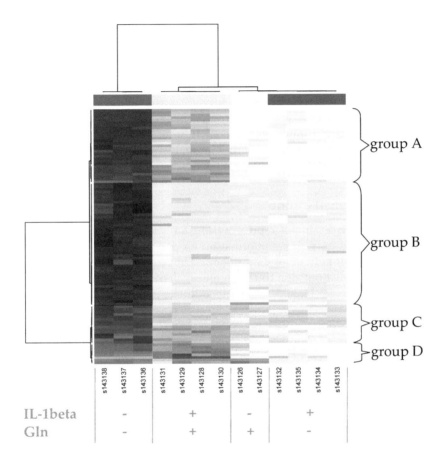

FIGURE 6.1: Heatmap of dataset GLUCO. Genes are in the rows and samples in the columns. The expression levels are indicated via a color coding. The shades of blue and red refer to the absolute expression levels; the darker the blue the lower the expression level, and the darker the red the higher the expression level. The dendrograms are based on Ward's clustering using the Pearson correlations between the sample or gene profiles.

the same section to use Ward's clustering on the Pearson correlations between the sample/gene profiles.

Heatmaps are sometimes used to show summarized values such as averages over replicates or fold changes between two groups. Although this approach might provide a better focus, one should realize that it ignores important information. As a heatmap can only show one color per gene and sample, it is impossible to show how variable the measurement across replicates is. When visualizing fold changes rather than absolute gene expression levels, the important information about the absolute abundance is lost.

Remember that a nice clustering of samples in a heatmap of some selection of genes is quite meaningless. In high-dimensional data you will always find something, but this might be due to multiple testing or overfitting (see Section 5.2). Even in a completely random dataset, for example, a heatmap will generally provide a perfect discrimination of the samples of two groups if it is based on 50 differentially expressed genes between these two groups.

6.1.2 Intensity plot

This plot is a simple representation for the expression levels of a single gene. While it only shows information for one gene, it has the advantage that it can show all individual data points. This is particularly useful after a gene has been singled out due to the significance of its alteration between samples and/or due to its biological relevance in the context of the biological experiment. The average group intensities are indicated at the y-axis to facilitate group comparisons (see Figure 6.2). Due to its clarity, this plot can be used both for interpretation purposes as well as in presentations.

The disadvantage of intensity plots is that it can only handle a very limited number of genes while a heatmap can easily show the expression profiles of 100 genes at a time. On the other hand, the plots are easier to interpret compared to the heatmap. As the reader first needs to translate the colors of a heatmap to the respective expression levels, its interpretation is more difficult and cumbersome. Also, because of the large amount of information, one may miss some interesting patterns like an extraordinary expression of a particular gene in a few samples in a heatmap.

Depending on the design of the study, some adaptations of this graph may be more appropriate. If the same subject has been measured repeatedly over time, it may be wise to plot time in the y-axis, to color by treatment and to link the measurements of the same subject with a line.

Example. Compare the intensity plots of Figure 6.2 with the heatmap of Figure 6.1. Figure 6.2 highlights four genes selected to represent each of the four gene groups clustered in the heatmap. The same patterns are much easier to observe, and the levels of differential expression are better appreciated. To summarize the patterns described in the previous section; the genes of groups A and B are down-regulated, while the genes of groups C and D are up-regulated in all three treatments compared to the control. In addi-

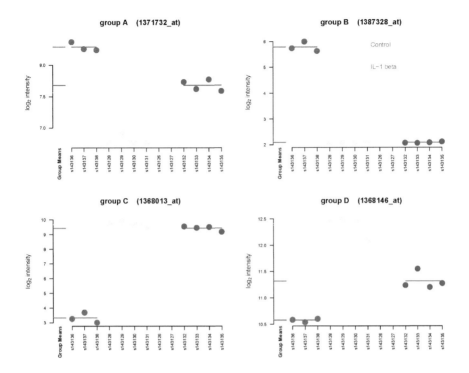

FIGURE 6.2: Intensity plots of four genes from the GLUCO dataset, showing the expression levels for all samples. Next to y-axis the average intensity per sample group is given.

tion, the genes of group A are down-regulated in the combined glucosamine/ IL1-β treatment. Finally, genes from group D are down-regulated by IL1-β treatment in the absence of glucosamine while genes from group C are not.

6.1.3 Gene list plot

A gene list plot visualizes the gene alterations for one or more groups. Each row represents a gene, generally together with its symbol and description. Each column visualizes the expression changes of a treatment compared to a reference, like a control treatment, by means of a barplot. Boxes to the left indicate down-regulation and boxes to the right up-regulation of a gene. To allow a better estimation of the size of the change, grey vertical boxes mark 2-fold change and, if necessary, extra grey lines indicate 4-fold change. The error bars, showing the pooled standard deviations, indicate the variation among the replicates in the given treatment group and the corresponding control.

In contrast to the heatmap and the intensity plot, gene list plots do not visualize raw data anymore. By calculating differences between two groups, and the variation on the estimate of this difference, the raw data are summarized into relevant characteristics. This reduction of information makes a full comprehension of the data difficult, but allows a better interpretation of the patterns of interest in the data. Indeed, gene alterations are much more appreciated as they have been calculated in advance, and this enables a more focused interpretation.

A gene list plot can be constructed in multiple ways to guide the reader of the plot. The names and descriptions of genes can easily be incorporated in the graph, and the variation of the estimated differences can be highlighted with error bars. Optionally, information on the absolute expression levels can be reintroduced via a color coding of the bars, for instance with three different colors to represent low, median and high expressed genes.

The genes can be ordered in different ways:

1. Ordered based on hierarchical clustering similarities to group genes with similar expression profiles across samples together.

2. Ordered based on significance to rank the most significant genes on top.

3. Ordered according to the changes in one particular group. This type of ordering is used in Figure 6.3.

As this plot is very data rich and focused on a crucial aspect like gene alterations, it helps to get a better understanding of the data. As the graph can be visually attractive, it may also be used for presentation purposes if the gene list is not too long (see Figure 6.3).

Example. The aim in the ONCO study is to differentiate genotoxic from general cytotoxic effects using transcriptomics. To this end, gene expression is measured on yeast after treatments with genotoxic compounds (methyl-methanesulfonate, bleomycin, cisplatin) and non-genotoxic compounds (ethanol and sodium chloride) together with a control, and differential expression is tested between the genotoxic and the non-genotoxic compounds. Figure 6.3 nicely summarizes the fold changes of all compounds vs. control for the 20 most significant annotated genes. Ethanol and cisplatin were only measured in two yeast samples, making it impossible to calculate deviations. The other compound treatments were measured in triplicates and have error bars.

The top 8 genes are up-regulated by non-genotoxic compounds, but not changed by genotoxic compounds. In contrast, some of the genes like TRF5 at the bottom of the graph are strongly down-regulated by non-genotoxic compounds while being up-regulated by the genotoxic compounds. The grey boxes and grey lines indicate that methyl-methanesulfonate and bleomycin up-regulation is more than 2-fold up-regulation, and the down-regulation of the two non-genotoxic compounds is even lower than 8-fold.

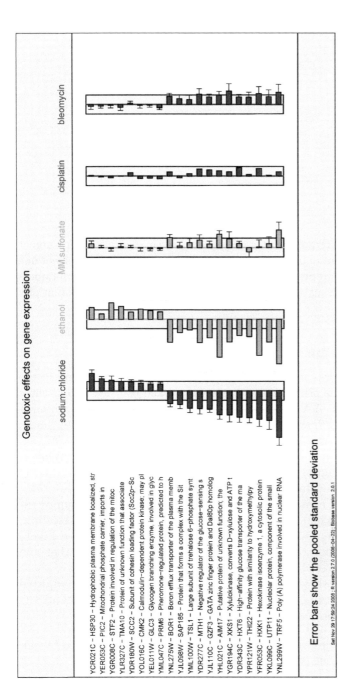

FIGURE 6.3: The 20 genes that are most differentially expressed between genotoxic and non-genotoxic compounds in the ONCO dataset. Each row represents a gene with its description and gene symbol at the beginning of the row while each column represents the gene expression changes between a sample and its corresponding control. Boxes to the left of the middle line indicate that a given gene is down-regulated, while boxes to the right depict up-regulation of a gene. To aid the observer of the graph in more accurately estimating the size of the change, the border of a grey vertical box underneath each gene box marks a 2-fold change. For larger gene changes an extra line indicating a 4-fold change is drawn. Furthermore, the pooled standard deviation for a given sample plus its corresponding control is indicated by error bars.

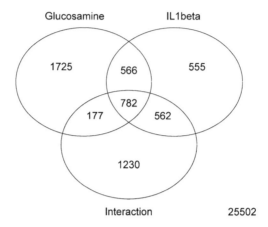

FIGURE 6.4: Venn diagram. The graph consists of three circles which depict three different groups or experiments. Where two or three circles intersect one another this area will represent overlap between the groups. The numbers inside the areas represent the number of genes that are, e.g., differentially expressed in a group.

6.1.4   Venn diagram

The Venn diagram was introduced by John Venn in 1880[232], and in the microarray context it is primarily used to compare and contrast two or three experiments or groups. A Venn diagram can highlight how similar or distinct groups are by showing how many genes with a certain pattern they share. As a graph should be easily readable and interpretable, Venn diagrams are typically used when comparing two or three groups or experiments. While it is possible to create diagrams for more than three groups, such graphs quickly become difficult to understand.

Typically, the graph consists of circles which each depict one group or experiment. The area where two or three circles overlap (the "intersection" between the groups) represents the overlap between the groups (see Figure 6.4). The numbers inside the areas represent the number of genes that are, e.g., differentially expressed in a group (vs. its control).

Example. Figure 6.4 shows the number of genes with significant unadjusted p-values (i.e., with a p-value $< 5\%$) for three hypotheses tested in the GLUCO dataset. The three hypothesis tests are

1. Glucosamine: genes that are significantly differentially expressed by glucosamine treatment (Gln + vs. Gln -).

2. IL1beta: genes that are significantly differentially expressed by IL1-β treatment (IL1-β + vs. IL1-β -).

3. Interaction: genes where the effect of glucosamine treatment (Gln + vs. Gln -) significantly depends on whether the samples received an IL1-β treatment (IL1-β +) or not (IL1-β -).

There are 25,502 genes that were never called significant. There are also 1,725 genes that were only affected by glucosamine treatment, while 555 genes were only affected by IL1-β treatment. There are 1,230 genes that had a significant glucosamine by IL1-β interaction, but were overall not changed by glucosamine nor by IL1-β treatment. These genes have no apparent overall glucosamine or IL1-β treatment effects because of the dependency between the two treatment effects. There are 566 genes that were significantly affected by both glucosamine and IL1-β treatment in an additive way (i.e., the effects of the two treatments on gene expression did not interact with each other). For 177 genes there was no overall IL1-β effect but a glucosamine-dependent IL1-β effect in addition to an overall glucosamine effect.

Depending on the research question, the researcher my want to pick a certain gene list using this Venn diagram overview (see Section 5.20 for a selection of four genes based on this Venn diagram).

6.1.5 Scatter plots

6.1.5.1 Volcano plot

In Section 5.5 we discussed different approaches for the detection of differentially expressed genes. The most commonly used are ranking genes by either the fold change (the effect size) between two groups or by the statistical significance estimated by an (unadjusted) t-test. The volcano plot is an effective and easy-to-interpret graph that summarizes both fold-change and t-test criteria (see Figure 6.5). It is a scatter-plot of the negative log10-transformed p-values from the gene-specific t-test (calculated as described in the next section) against the log2 fold change. For every gene that was measured on the microarray, a dot in the graph represents its fold change and t-statistic. One of the main motivations to use volcano plots routinely is the wish to discount significant genes with misleadingly small variances and fold changes[233].

The genes at the top of the graph are statistically the most significant and genes at the left and right side of the graph have the largest fold-changes. Accordingly, the genes in the upper-right and in the upper-left corners are the most interesting genes, as they show both a strong effect as well as high statistical significance. A simple but very informative approach is therefore to highlight the genes above a horizontal threshold line and/or outside a pair of vertical threshold lines by their respective gene symbols in text. A statistically more sound alternative to this ad-hoc approach is the two-dimensional local FDR[233]. This methodology allows to objectively define regions of interest in a volcano plot (i.e., the upper-right and upper-left regions) by incorporating the standard error and the t-statistic into one adjusted p-value for every gene.

For an audience used to looking at volcano plots, such a graph can be included into a presentation. Otherwise it should mainly be used for data analysis purposes, as it has a high information density.

To enhance readability of the plot, the genes close to the center of the coordinate system – the genes one is typically not interested in – are commonly visualized by a smooth scatter plot that highlights areas of many genes. The corresponding shading is typically darkest in the direct vicinity of the center of the coordinate system as the biological experiment should have been controlled in such a way that only few genes actually change between samples.

Example. Figure 6.5 is a volcano plot of the BCR/ABL vs. NEG comparison using the B-cell samples of the ALL data. Interestingly, the top three probesets all measure the same mRNA transcript, i.e., the gene ABL1. They are by far the most significant and, although they do not have the largest absolute difference, their fold changes are also among the largest observed. This observation makes sense as the main difference between the two groups being compared is a mutation in this particular ABL gene.

6.1.5.2   MA plot

In essence the MA plot visualizes the average signal intensities for all genes between two samples vs. the differences in intensities for all genes between the two samples or two groups. The abbreviation MA stands for plotting the average (A) of the log intensities vs. the differences in average intensities (minus, M).

There are two other commonly used terms for this type of plot: the MA plot is also called "ratio intensity plot" or "Bland-Altman plot." Originally this plot was intended to visualize the dependency of the fold change between two groups on the intensity for all genes when looking at raw microarray data.

As the biological experiment underlying the microarray experiment should be performed in such a way that the majority of genes are not affected, one expects to see most points (genes) close to the y-axis, since log(1) is 0.

Especially in the field of dual-color microarrays where one sample is labeled with one dye and another sample is labeled with a different dye, dye-specific artifacts were visualized with the MA plot (see Figure 6.6). Thus it is being used as a diagnostic plot to check the performance of the applied normalization procedure. As it is typically known only to researchers familiar with microarray technology, this graph is rarely used in presentations as it looks at technical aspects rather than presenting biological results.

Example. Figure 6.6 is an MA plot of the BCR/ABL vs. NEG comparison using the B-cell samples of the ALL data. As expected, most genes are not differentially expressed. The genes with the highest fold changes are highlighted, and based on the gene symbols one can easily find a consistency with the x-axis of the volcano plot in Figure 6.5. The MA plot is not indicating the biologically more important genes, such as the gene ABL1. This illustrates

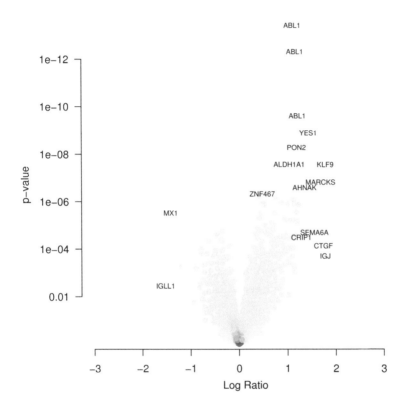

FIGURE 6.5: Volcano plot. Every gene is represented by a dot in the graph. The x-axis represents the fold change and the y-axis the t-statistic. To enhance readability of the plot, the genes close to the center of the coordinate sytem are only visualized by a smooth scatter plot whereby the darkness of the blue color correlates with the number of genes in a particular area. The more genes, the darker the color. Genes most distant from the center of the coordinate system are represented by their official gene symbols.

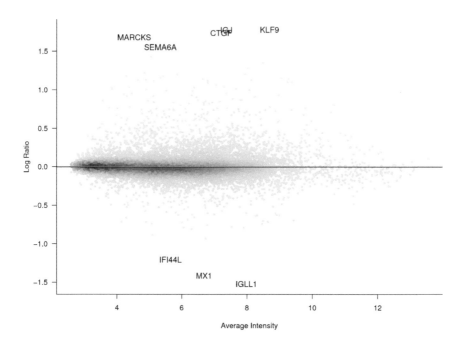

FIGURE 6.6: MA plot. The MA plot shows differences in average intensities (M) on the y-axis and average intensities (A) on the x-axis for each gene (dot) on the plot.

that the merits of an MA plot are primarily to explore technical quality of the samples rather than to discover biological findings.

6.1.5.3　Scatter plots for high-dimensional data

The classical scatter plot as shown in Section 4.2.3.2 should not be used anymore. We recommend the use of MA plots for assessing the quality of a normalization or the use of a volcano plot for looking at raw effects as the graph will visualize both the fold change as well as the statistical significance between samples or sample groups. Furthermore, as high-dimensional data such as microarray data have very many points, e.g., thousands of genes, binning and visualizing the density of non-informative points as done in smooth scatter plots aids in interpreting a graph. It also reduces the file size when making graphs in a vector format like postscript or pdf and accordingly reduces rendering time in viewers such as Adobe Acrobat.

Example. Figure 6.7 shows the relationship between variance and mean across all samples for all genes. The parabolic relationship between mean and

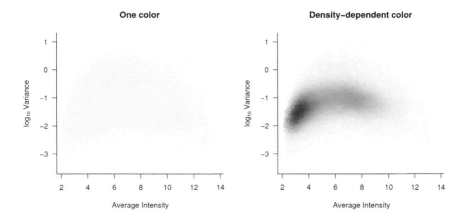

FIGURE 6.7: Smooth scatter plot. To identify patterns relating to the density of data points in a given area of the plot, the smooth scatter plot only depicts single data points in areas that are not densely populated. Areas with very many data points are transformed into a color shading whereby a darker color visualizes more data points as compared to a lighter color. Here we plotted the variance vs. the mean for all genes across all samples of the GLUCO dataset (each dot represents a gene). The parabolic relationship between mean and variance becomes more clear when using a density-dependent coloring.

variance becomes more clear when using a density-dependent coloring. This relation is due to the dynamic range of microarrays, which can vary across experiments but is typically between 2 and 2^{16} on raw scale. The probes of genes with very high expression levels therefore get saturated towards 14-16 on a log_2 scale. Genes with average expression levels have the most potential to vary within the boundaries of this dynamic range.

6.1.6 Histogram

A histogram visualizes how frequently certain values of a (continuous) variable occur across the samples. It first splits up the continuous variable into categories and subsequently calculates the proportion of samples that fall into each category. A histogram therefore shows the frequency distribution of the variable.

In a microarray setting, histograms of the p-values of tests for differential expression (see Section 5.5) are of crucial importance. They are simple, easy to interpret and contain important information. A histogram shows whether there is a signal in the data or not; i.e., whether there are genes being differentially expressed or not. Furthermore, it can also be used to get a feeling of how many genes are expected to be truly differentially expressed. This con-

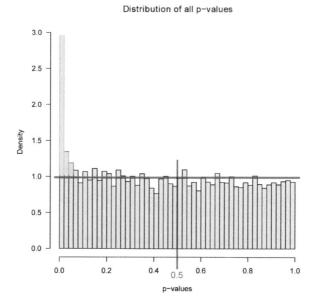

FIGURE 6.8: Histogram of p-values for the t-test testing differential expression between ABL/ALL and NEG using the ALL dataset. The red line visualizes a uniform distribution, which reflects a distribution of p-values. The barparts highlighted by the green surface indicates the number of genes that are expected to be truly differentially expressed.

cept is used by Storey[234] to estimation of expected number of false positives in the calculation of the positive false discovery rate (see Section 5.5.3.2.3).

A histogram of p-values of a dataset where no gene is differentially expressed will look at a flat line, visualizing a so-called "*uniform distribution.*" Because there is no signal in the data, any value between 0 and 1 is equally likely, resulting in a flat distribution. Just a side note, about 5% of the p-values will be less than 0.05, a logical result of the definition of a significance test (this is being discussed in the context of multiple testing in Section 5.5.3.1). In other words, if the p-values of a test have a flat distribution, one should be careful in looking at the top significant genes, even after correction for multiple testing. The uniform distribution of the p-values namely indicates that presumably only false positives will be found.

Example. Figure 6.8 shows the p-value distribution of the test for differential expression between ABL/ALL and NEG using the ALL dataset. Clearly the distribution does not follow a uniform distribution, indicated by the red line. Compared to this flat distribution, there is an enrichment of genes with p-values close to zero (bar parts highlighted in green). In other words, there are more significant genes than expected just by chance. This is the best indication that there is some signal in the data. Unfortunately, it cannot be used to identify which genes are the true and which are the false positives. Indeed, the bar at the extreme left of the histogram is not entirely green (the expected true positives) but also consists out of one third of blue (the false positives). Luckily, the extremely significant genes are much more likely true positives, but this is not universally valid.

6.1.7 Box plot

Box plots or box-and-whisker plots[1] were invented in 1977 by the American statistician John Tukey to visualize robust statistics. They essentially visualize the distribution of the data (in the y-axis) per category or sample (in the x-axis, see Figure 6.9).

In the microarray context, box plots indicate the intensities of 50% of the transcripts around the median, whereby the lower end of the box refers to the intensity of the mRNA at the 25 percentile (the lower quartile) and the upper-end of the box indicates the intensity at the 75% (the upper-quartile). Furthermore these plots also show transcripts that have a particularly low or high intensity for a sample. When doing microarray data analysis these plots are often drawn to visualize raw probe intensities (before summarization), as they can help in identifying possible technical outlier arrays (see Figure 6.9). A close relative of the box plot, the violin plot, is described below.

6.1.8 Violin plot

The violin plot is a boxplot modification that provides a more detailed view of the shape of the distribution (see Figure 6.10). In other words, this plot preserves the features of the density plot (see below) and combines it with elements of the box plot. Initially a box plot is drawn. A rotated kernel density plot is plotted twice on top of the box plot. The last step is overlaying a line indicating the median of the dataset. To enhance the visualization of the density plot, the area between both kernel density plots is colored. If the data have a bimodal distribution (i.e., have two distinct normal distributions), the plots get the shape of a violin, hence the name. Violin plots tend to be easily interpretable for a limited number of samples[235].

---

[1]When applying box plots to check for outlier arrays, one typically looks primarily at the relative position of the boxes (the intermediate 50%) between the samples. When using this kind of plot for a different purpose, one needs to be aware of the different ways how the end of the whiskers are defined. While they can depict the minimum and the maximum of all the data, they could also represent alternative values, e.g., one standard deviation above and below the mean of the data. Therefore, one should always describe in the figure caption what the whiskers represent.

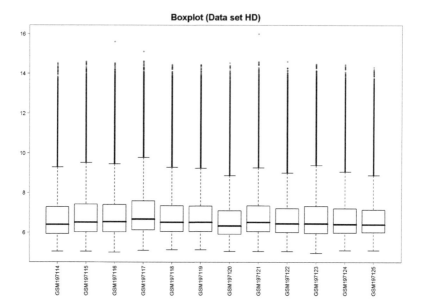

FIGURE 6.9: Box plot of dataset HD. Looking at the box plots of all samples in the dataset one cannot identify a sample that shows a particularly different intensity distribution when compared to the other samples in the experiment. Only sample GSM197120 seems to show somewhat lower intensities overall, but typically this will be corrected by the normalization step.

6.1.9   Density plot

The density plot is a diagnostic plot that visualizes data distributions. In a microarray context, it is typically used to identify outlier samples. The graph shows how frequent a certain intensity was measured for a sample. If there are problems with the sample quality or, e.g., the microarray hybridization, the curve for this sample is typically more steep toward low intensities and will have a smaller shoulder of higher gene expression levels (see Figure 6.11).

Example. Figure 6.11 shows the distribution of all samples of the GLUCO dataset. The density lines are colored by treatment. There are no clear outliers, although two samples are a little off. As these two samples have different colors (blue and green), they do not belong to the same treatment.

6.1.10   Dendrogram

A dendrogram (from Greek dendron "tree," gramma "drawing") is a tree diagram frequently used to illustrate the arrangement of the clusters produced by a clustering algorithm (see Section 5.4.2). Dendrograms are often used to

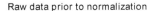

Raw data prior to normalization

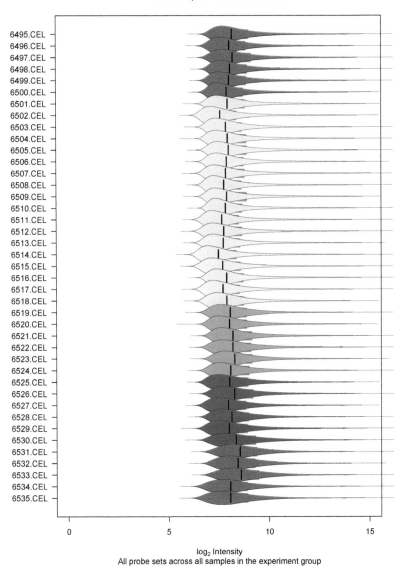

FIGURE 6.10: Violin plot. This plot combines a box plot with a density plot. In the example above the experiment had seven treatment groups. Each group has a different color. In each row one sample is visualized. The median intensity for the sample is shown by a vertical line.

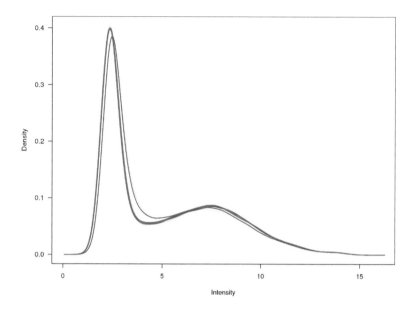

FIGURE 6.11: Density plot of dataset GLUCO. Each curve in the graph depicts data for one sample. The y-axis denotes the frequency how often mRNAs
have a certain abundance as shown by the x-axis. The x-axis represents the
mRNA abundance by giving the range of the measured intensities.

illustrate the clustering of genes or of samples. They are often used to provide
more structure to heatmaps (see Section 6.1.1).

The terminal branches (called the leaves) of the dendrogram are linked with
the individual genes or samples, and the length of a branch is typically proportional to the pairwise distance between the clusters. Section 5.4.2 contains
more details and recommendations such as which distance or linkage function
to use to construct the dendrogram.

Example. The dendrogram in Figure 6.12 shows the clustering of the samples of the ALL data. The samples fall in two predominant clusters which
correspond perfectly with the type of the ALL cell, namely B- or T-cells.

6.1.11   Pathways with gene expression

Figure 6.13 shows the same results, but then the results are visualized in a
pathway so that their interconnections can be studied more closely.

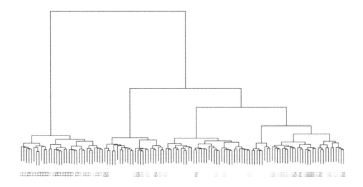

FIGURE 6.12: Dendrogram of the ALL data based on Pearson distance and Ward's clustering. The labels are colored by type of ALL cell; labels of B-cells are colored green and labels of T-cells are colored red.

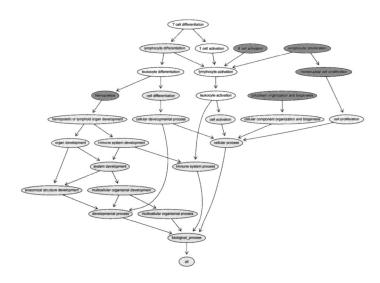

FIGURE 6.13: GO pathways containing the gene sets that were most significantly up-regulated or down-regulated in T-cells compared to B-cells in the ALL dataset. The significance was obtained using GSA. The same figure is shown in more detail in the pathway analysis section in Chapter 5.

6.1.12   Figures for publication

Until now we have made primarily a distinction between graphs that are intended to be used for data analysis and those that are supposed to be shown during a presentation. There is however another important area: the generation of figures that are to be included in publications.

It is impossible to visualize all information in a single graph. This is especially true for microarray data. Therefore one should try to limit the message of a figure to one or two observations. All the details of the graph should contribute to these one or two observations only to increase the clarity of the figure.

While each journal has its own set of rules on how to prepare figures, the following rules typically apply:

- For figures including lines or curves, a rule of thumb is to use lines that have a thickness between 1 and 2 points (a thickness between 0.5 and 1 points is acceptable for some journals as well). If color is not an option when needing to include multiple lines in a figure, either symbols or dashed lines. Furthermore, be sure to provide labels for all axes and all lines.

- While text is often more easily readable with serif fonts (e.g., Times New Roman or Palatino Linotype), many journals prefer to receive figures that make use of sans serif fonts (e.g., Helvetica or Arial), as underline the clarity of a figure. Since it is often necessary to either enlarge or shrink figures for the final article, the figure should preferably use only one or two font sizes of similar size. When text is placed inside the figure on a dark or textured background, the readability is enhanced when placing a white area behind the text. Be careful not to use multiple fonts. The fonts that are used should be embedded into the electronic file (e.g., pdf).

- One also should keep in mind the extra charges for color figures. Even though color is an excellent tool for helping to explain complex data, sometimes observations can be demonstrated equally clearly using black and white figures (e.g., box plots, MA plots, intensity plots).

- Presentations do not require figures to have a higher resolution than 72 dpi. For printed publications, however, photos should have 300 dpi and black and white line art (if not created in a vector format) should be 1,200 dpi. Lower resolutions will result in pixelled images that are often rejected by publishers. If at all possible, it is preferred to generate figures in a vectorized format (e.g., Adobe Illustrator, Adobe Freehand, Inkscape, etc.) over pixel-oriented formats.

6.2   Biological interpretation

Besides using various statistical tools and bioinformatics software, the scientist still needs to put the experimental results into the biological context of the system or process that was studied. As this part of the data analysis is the most subjective part of the complete process – the outcome depends heavily on the prior knowledge of the scientist – it is impossible to give suggestions that will be applicable for all microarray experiments.

However, below are a number of tools and data sources that provide starting points for investigating microarray findings further. They provide links to gene annotation, to knowledge about processes and diseases in which genes were reported to play or role, or point towards other tools such as text mining applications.

6.2.1   Important databases

All molecular databases have a basic annotation unit. This can be a gene, a transcript, a peptide, etc. A problem arises when one would like to search for information on something smaller than this basic unit, e.g., exon information in a database that uses transcripts as its basic unit.

6.2.1.1   Entrez Gene

Entrez Gene is the freely accessible online repository for gene-specific information at the National Center for Biotechnology Information (NCBI), a division of the National Library of Medicine (NLM), located on the campus of the US National Institutes of Health (NIH) in Bethesda, MD, USA. While it does not contain all known or predicted genes, it includes records from genomes that have been completely sequenced. It also covers gene information on genomes that are going to be fully analysed. Entrez Gene pulls together gene-centered authoritative information from a number of resources: results of both curation and automated analyses are integrated, including data from NCBI's Reference Sequence project (RefSeq)[22], information from collaborating model organism databases and data from other databases within NCBI, like the Gene References into Function (GeneRIFs)[22].

The following information can be retrieved for a record: nomenclature, chromosomal localization, gene products and their attributes (e.g., protein interactions), associated markers, phenotypes, interactions, as well as links to citations, sequences, variation details, maps, expression reports, homologs, protein domain content and external databases.

6.2.1.2   NetAffx

The NetAffx website is a resource maintained by Affymetrix. Especially when using their arrays in the way as suggested by the manufacturer, this site is invaluable as it provides all relevant information about the array content, the probeset definitions, the consensus sequences that the probesets were designed from, the most current annotation for the original probeset definitions, and much more.

6.2.1.3   OMIM

For microarray experiments that are run in the context of a human disease, OMIM is a very useful database as it provides linking information between human genes and all known mendelian disorders. As it focuses on human diseases that are known to have a genetic component, it is an ideal starting point to search for links between differentially expressed genes and potentially related medical conditions. Information on more than 12,000 genes are currently contained in the daily updated database. The database has been available on the Internet already since 1987 and in 1995, OMIM was adjusted to the requirements of the World Wide Web by the National Center for Biotechnology Information.

6.2.2   Text mining

With the rapid growth of biomedical literature, it becomes impossible to read all relevant literature in a particular scientific field. Therefore a number of commercial and academic groups are investigating technologies to mine the literature for, e.g., connections between genes and diseases. Text mining attempts to define relationships between biological entities when they are discovered in multiple articles. For microarray data, there are approaches that search the literature for co-occurrences of genes that were found to be differentially expressed. Based on these co-occurrences, concepts are extracted and visualized. The aim is to identify themes similar to the identification of significantly affected gene sets in pathway analysis. Here, however, the biological interpretation is based on co-occurrences of genes in biomedical articles and not based on, e.g., an over-representation of gene sets with a small p-value.

6.2.3   Data integration

6.2.3.1   Data from multiple molecular screenings

The interpretation of gene expression data can be further enhanced when profiling data of other molecular screenings of the same sample is available. However, depending on the type of data, the correlations with mRNA measurements will be less. For example, while there have been a few reports on the limited correlation between changes in mRNA abundance and changes in protein abundance (as measured by mass spectrometry based technologies), a recent publication has highlighted that the correlation seems larger when comparing mass spectrometry data with exon-level microarray data. In this approach individual peptides are mapped to corresponding exons via the genomic DNA sequence[236].

6.2.3.2 Systems biology

The aim of this scientific discipline is to build predictive probabilistic networks. The basis for building such networks is knowledge from different levels. Identifying genes that are co-regulated is one source of information. Others include assumptions like "associated proteins are likely to be involved in the same biological process or carry out similar functions." Whole genome gene expression studies are currently often used for defining co-regulated genes. "Guilt by association" is a term that is often mentioned in the context of systems biology. It refers to the concept of inferring a gene function from being involved in similar traits. In other words, knowing the effect of manipulating one gene with a known function can help in deducing the gene function from a not characterized gene if this gene leads to the same effect. One goal of systems biology approaches is to model human diseases. Good models truly characterizing a disease could prove to be invaluable tools for the identification of genes that are most relevant for the disease. In turn and if drugable, they could become targets for screening for compounds modulating the activity of the proteins encoded by these key genes.

6.2.4 RTqPCR verification

While there are still a number of reviewers and journals that ask researchers to confirm the results of their microarray experiments with RTqPCR, we tend to discourage scientists to spend the money and time for such experiments.

One main reason is the high confirmation rate. We routinely have well above 90% confirmation when the RTqPCR primers and probes were selected to cover the same region as covered by the probes on the Affymetrix arrays. Cases where the confirmation did not work often turned out to be caused by poor primer PCR performance, bad annotation of the probeset, etc. Furthermore, since we have switched to making use of the alternative CDF mappings (see Section 2.2.2), the success rate has increased even further.

While authors of the MAQC study have already reported that RTqPCR data is not always as accurate as microarray data for various reasons (e.g., choice of housekeeping genes that can change at times, better pre-processing techniques for microarrays due to the availability of robust techniques such as quantile normalization), we share the experience of Jenny Drnevich that he has communicated in the forum of the open-source gene expression analysis software Bioconductor. He reports about a client that has the experience from a client who had confirmed the down-regulation of gene knocked out in a transgenic animal via quantitative RTqPCR. During a subsequent Affymetrix microarray experiment, he actually saw an up-regulation of the gene that was supposed to be knocked out. How could this be?

It turned out that the gene knock-out was actually only a partial deletion of two exons of a gene. As the Affymetrix probes were covering a different region they correctly detected a truncated transcript that was still expressed. Furthermore, as the truncated transcript resulted in a non-functional protein, the organism tried to compensate the lack of functional protein by up-regulating the transcription of the gene to make more protein. In other words, both technologies were correct, but measured different things. Therefore, one should be very careful with confirmatory RTqPCR experiments.

However, we certainly do recommend the use of quantitative RTqPCR for new experimental samples where one wants to save the costs of running full microarrays and rather wants to focus on confirming the findings of a previous array experiment.

6.3   Data publishing

Most journals require the deposition of raw microarray data in community-endorsed public repositories prior to manuscript submission. The paper then needs to refer to the corresponding accession number. The goal of this policy is to allow other researchers to reanalyze the data. This is especially necessary as it is often impossible for a reviewer to assess the conclusions that are drawn in a manuscript. However, by depositing the data into a publicly accessible repository, it is ensured that others working in the same scientific field, can reanalyze the information and compare it with their own findings.

There are two main data repositories (GEO and ArrayExpress), which are introduced next. While they have slightly different goals and different approaches, they both are suitable and accepted by all major publishers. As a simplified summary one could state that GEO focuses more on data storage, and ArrayExpress more on data access.

6.3.1 ArrayExpress

ArrayExpress[237] is a MIAME[2] (minimum information about a microarray experiment)-complient public database for gene expression data. The generic design of the database does not limit it to one or few microarray technology platforms or species. It is sponsored by and hosted at the European Bioinformatics Institute (EBI) and collects data since February of 2002.

The major goals of ArrayExpress are to serve as a repository for data that are accompanying a publication, provide and share scientific data for future reanalyses or meta-analyses in a standardized form and to stimulate an open exchange of all parameters that are relevant for a microarray experiment (e.g., platform design, array content, lab protocols, etc.).

Furthermore, data on catalogue arrays from Affymetrix and Agilent are retrieved on a weekly basics from Gene Expression Omnibus. If they pass the manual curation process, they are uploaded into ArrayExpress.

6.3.2 Gene expression omnibus

Similar to ArrayExpress, Gene Expression Omnibus (GEO)[238] is also a repository for gene expression data or other molecular abundance information at the National Center for Biotechnology Information (NCBI). The database was established in 2000 and equally supports the submission of data according to the MIAME guidelines. The curated data is made available via an online interface that allows the user to query or browse through the stored gene expression data and retrieve the data and, where available, the raw data such as .CEL files. This Web interface can also be used to visualize the stored data on an individual gene level or on the level of an entire study.

Compared to ArrayExpress, GEO also has a flexible design that allows the user to submit data in different styles (e.g., unprocessed or processed data in a MIAME-supportive infrastructure) for different platforms and species. In contrast to the goals of ArrayExpress as mentioned above, the primary role of GEO is data archiving by means of a robust and versatile data repository. This is underlined by the design decision to not enforce any restrictions or login requirements for the storing and distribution of the contained data. As a consequence, GEO has grown to be the largest public gene expression database. While ArrayExpress contains microarray data only, GEO also contains data from other molecular profiling technologies, such as SAGE or mass spectrometry peptide data.

---

[2]This standard represents a recommendation of the Microarray Gene Expression Data (MGED) society (http://www.mged.org). By presenting the scientist with a list of information that is necessary to describe and replicate an experiment, the MGED society aims to make the submission of microarray data an integral part of a publication submission process.

6.4   Reproducible research

Reproducibility is a key issue in microarray research as it is linked to false positives. Even when lab work and statistical analysis were carried out perfectly, microarray experiments may generate false results if the sample size was not large enough due to the curse of dimensionality (see Section 5.2). Reproducibility will unmask these false positives as only real results are reproducible.

Reproducibility is a cornerstone of good scientific practice. The idea behind reproducible research is that anything in a scientific paper should be reproducible by the reader. In practice, this is often difficult because the published details of the, generally complex, used methodologies are too scanty for anyone to reproduce it[239]. This lack of detail is not entirely the author's fault but often also due to limitations of the journal with respect to the available number of pages. This regularly does not leave room for full explanations. Fortunately the Internet, with no page pressures, comes to the rescue. Many scientific papers nowadays already point to supplementary materials available on-line. This allows to make the used software code and the data publicly available.

Ideally, the entire study should be reproducible. For many aspects of the study, including all the sampling and lab work, this is difficult or impossible. This in contrast to all the steps of the data analysis that should in principle be perfectly reproducible. It only takes will and knowledge to do it[239]. To allow reproducibility of the results, a good documentation of the followed steps is however necessary. Unfortunately, many data analysts regularly perform a "point&click" in a menu-driven software tool, or a "copy/paste" by hand without documentation. One such step makes the analysis already unreproducible by third parties. There are nowadays tools available to create dynamic documents: entire reports that include all data analysis steps and all code used to make plots. These documents are perfectly reproducible, transparent and dynamic. They are described in more detail in Section 8.2.

Chapter 7

Pharmaceutical R&D

This chapter provides an overview of the different application areas of gene expression profiling in pharmaceutical research. While this chapter may not be interesting for all readers, we believe that it contains concepts that are applicable in general. The following sections describe how gene expression profiling experiments are carried out on a daily basis in a Pharma R&D setting. Putting the different applications into the entire context of how a drug is developed will also demonstrate how widely gene expression profiling is applied throughout the whole drug discovery and development process (see Figure 7.1).

7.1 The need for early indications

The costs for developing novel medical products are high and continuously rising. The development of innovative drugs has become more risky as well, particularly for less common indications or for diseases that are more prevalent in Third-World countries. As a result, the number of new molecular entities submitted to the US Food and Drug Administration (FDA) has steadily declined during the last decades. Hand in hand with these declines go decreases in regulatory approvals.

A major driver for the increasing costs is the high failure rate in phase III clinical trials where most failures are due to a lack of efficacy (the new drug was not more effective than the current standard care). This explains the need to develop markers that are indicative of efficacy to allow earlier decision making. Phase II trials may already bear some proof of compound efficacy, but generally have too small sample sizes for drawing strong conclusions.

Efficacy markers may increase the success rate of clinical trials, especially when dealing with first-in-class compounds that work via novel mechanisms of action. These compounds have the drawback that there is little to no prior experience with suitable clinical endpoints, they are more often failing in phase III trials. Exploratory profiling technologies have a high potential to lead to efficacy markers. They do not need to be fully understood yet, particularly for novel mechanisms.

Target Identification and Validation
Disease Model Characterization
Identification of Disease Pathways and Mechanisms
Disease Marker Identification / Disease Subgrouping
Compound Profiling
Dose-Response Markers
Patient classification
Efficacy Markers
Outcome Prognosis
Toxicogenomics

FIGURE 7.1: Gene expression profiling in the context of Pharma R&D. Flowchart highlighting the main stages of the drug discovery and development process. The research domains where microarray technology is used are shown below the flowchart.

Especially the development of new biomarkers is thought to be aided by modern microarray technologies. Because of their high-content nature, they offer the possibility to provide a set of marker genes vs. single markers for a given condition.

Biomarkers can be highly valuable, e.g., in the context of personalized medicine where they provide doctors with the right tools to individualize therapies by identifying the patients that are likely to respond to a given drug (see Section 7.4.2). As such response markers are likely to originate from preclinical disease models, they in turn will require better predictive disease models (see Section 7.3.5). The increasingly popular field of biomarker research has motivated many pharmaceutical companies to seriously consider co-development of their candidate drugs together with compendium diagnostics as the most desirable way forward.

7.2 Critical path initiative

In March 2004 the FDA started an effort to try to modernize the scientific process by which potential drugs become medical products. This effort has been referred to as the Critical Path Initiative[240]. One of the thoughts behind this initiative was to bring novel scientific advances in drug discovery such as microarray technology more formally into the development of new drugs. Many consider some of the scientific tools that are still being used for the assessment of drug safety and efficacy as outdated. This leads to a

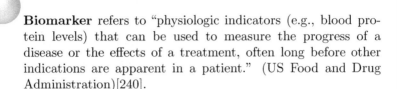

Biomarker refers to "physiologic indicators (e.g., blood protein levels) that can be used to measure the progress of a disease or the effects of a treatment, often long before other indications are apparent in a patient." (US Food and Drug Administration)[240].

BioBox 7.1: Biomarker

desire to benefit from novel developments in areas such as high content gene expression profiling or bioinformatics.

As microarray technology was classified by the FDA's Critical Path Initiative as a technology with "vast potential" for advancing medical product development and personalized medicine through the identification of biomarkers (see BioBox 7.1), the FDA started a major involvement in the development of standards that would eventually allow the use of microarray data in regulatory decision-making. Together with equipment suppliers, the FDA created a project consortium (called the "Microarray Quality Control" or "MAQC") to jointly prepare standards for the assessment of a laboratory's proficiency in carying out microarray experiments. Sections 2.5.1 and 5.5.2.1 discuss some of the conclusions from the first study conducted by the MAQC consortium. The second MAQC study is focusing on "best practices" for the development and validation of microarray-based predictive models. The results from this study, expected in April-June 2009, will evaluate the findings based on three categories of criteria for assessing the performance of predictive model: prediction performance, robustness of the signature, and mechanistic relevance of the signature, in decreasing order of importance for classification purposes.

A major breakthrough for the use of microarray technology was the clearance of the 3000Dx system[1] by the FDA in January 2005. This helped to establish the Affymetrix system as a platform for studies related to the registration of a new drug or for testing patient samples for diagnostic purposes. The system was approved as an in vitro diagnostic product in conjunction with the Roche Diagnostics AmpliChip®[2] CYP450 Test. This test detects genetic variations in the Cytochrome P450 2D6 and 2C19 genes. Identifying

---

[1]The GeneChip System 3000Dx (GCS3000Dx) is the first microarray instrumentation system for molecular diagnostic laboratories which has received US FDA (510k) approval. The system consists of the FS450Dx Fluidics Station for hybridization, washing and staining of the arrays, the GCS3000Dx Scanner with AutoLoader Dx for the automatic scanning of up to 48 microarrays, and a Workstation with GCOSDx software which controls the operation of the fluidics station and the scanner.

[2]Registered trademark of Roche Diagnostics, Indianapolis, Indiana.

variations in these genes can have a major role in predicting how a patient will metabolize many of the widely prescribed drugs. Drug metabolizing enzymes have an influence on drug efficacy and can lead to adverse drug reactions. Therefore the test leads to a prediction whether a given patient will likely be a poor, intermediate, extensive, or ultra-rapid metabolizer. As this test analyzes a patient's Cytochrome P450 2D6 and 2C19 genotypes, it is a test looking into the characteristics of a patient DNA and not the mRNA.

The FDA clearance of the Affymetrix system gave scientific credibility to the platform technology in a more easily controlable setting as the approved assay looks at the more stable DNA. But shortly after a second generation update of the GCS3000Dx system focusing especially on the use of the system as a diagnostic tool was launched in December 2007, the system was also cleared for gene expression profiling at the end of July 2008. The first approved expression-based diagnostic test of the Affymetrix system is a test that will aid in identifying the tissue origin of tumors that are hard to identify with conventional approaches. This approval is a major step forward in the context of the Critical Path Initiative as the system moves beyond a tool for biomarker discovery. Now it becomes possible to use the platform also to validate and apply the discovered biomarkers on the same system. It is also a practical step forward as it is not necessary anymore to spend efforts in converting the findings of the discovery phase into a validated marker on a different platform, e.g., RTqPCR assays (see BioBox 3.8). This used to be difficult at times as the measurement technology typically relies on other probes than the ones present on the microarrays. In other words, the realization of the aim of the Critical Path Initiative to bring novel advances in science into the development of new drugs has come a step closer.

7.3   Drug discovery

7.3.1   Differences between normal and diseased tissues

A classical application for gene expression microarrays is the comparison of normal tissue to diseased tissue. Here, scientists wish to identify differences that are consistently observed in the diseased tissue. As these differentially abundant mRNA molecules encode proteins, knowledge about the molecular function and the biological processes in which the encoded proteins are active, often provides the scientist with mechanistic information about the disease.

For example, in the study by Aerssens et al.[85] the researchers were interested in differences that could be observed in a specific colon tissue (biopsies from sigmoid colon mucosa) between 25 healthy volunteers and 36 patients suffering from Irritable Bowel Syndrome (IBS). Using Affymetrix Human Genome U133 Plus 2.0 arrays it was investigated whether potential differences could serve as a molecular biomarker for IBS and whether identifiable alterations could enhance our understanding of this disease. A set of 12 genes of which some are described to be involved in the host mucosal immune response to microbial pathogens were identified and confirmed by RTqPCR. The study also underlined the quality of the microarray technology as it was striking to see that the obtained gene expression profiles from a given patient were very stable for samples taken from different sites within the colon and on repeat biopsy after approximately three months.

7.3.2   Disease subclass discovery

Besides identifying the differences in diseased tissues, another field of research relevant to the pharmaceutical research is the investigation of the homogeneity of a population currently classified as suffering from the same disease. In oncology, one could try to, e.g., distinguish early stages of a certain cancer type from late stage cancer, or to identify tumor origin tissue type in metastasis. Another oncology-related example for this application is the low number of people responding to a targeted therapy. One potential reason for this variable behaviour could be explained by the presence of unknown subgroups in such a population. If one were to be able to subclassify a disease population further and correlate one of the classes with a characteristic that is linked to response to a treatment, more personalized medicine becomes a reality. As we have discussed in Section 1.2.1, the design of correlational studies is less controlled compared to experimental studies. Therefore they need many more samples to provide convincing evidence as they do not identify the cause of possible subgroupings.

Example. Dataset ALL is an example for such an application (see Section 1.3.1). In this study bone marrow samples, peripheral blood or both from 33 patients with T-cell acute lymphocytic leukemia (T-ALL) were profiled to search for distinct subsets of patients. Using an unsupervised hierarchical clustering technique on a set of 313 differentially expressed genes, two subgroups were identified. Further investigation identified a correlation of the groups with the degree of T-cell differentiation but which was lacking a correlation with clinical outcome. However, focusing on a limited number of genes, the authors of the study showed a predictive value of this gene set related to response to therapy and to survival.

7.3.3   Identification of molecular targets

Defining alterations between diseased patients and healthy controls lies at the basis of qualifying proteins as potential drug targets for medical intervention. A common application of gene expression arrays is to identify differential gene expression between a group of unmedicated patients and a control group. It is hypothesized that any observed alteration has the potential to be correlated with the disease state, either directly as causing the disease or indirectly as a consequence whereby the organism tries to adjust for an alteration.

It is, however, often quite difficult to disentangle causative genes from non-causative genes when looking at differences between diseased patients and healthy controls. Although non-causative genes appear less relevant at first sight, they might still influence disease progression. Depending on what genes were altered downstream of the causative genes, one might be able to utilize the information about downstream alterations to subclassify patients which in turn will have influence on the selection of a suitable treatment.

7.3.4   Profiling for molecular characterization

When designing an assay to identify active compounds or when designing models to simulate aspects of a disease, it is essential that the situation in the cells of a diseased organ or tissue are resembled as closely as possible. While cells from a cell line cultivated in the laboratory are even more distant than tissues from an animal model (they do not typically reflect the complex microenvironment in which a cell is located in a multicellular organism), it is desireable to know as much as possible about the elements that are indeed reflecting the situation in a patient's organ (such as the presence of a receptor that is to be modulated by a compound). But also those characteristics that are dissimilar are important to identify as they reflect the limitations of the model (such as genes that are active due to the fact that most cell lines are tumor derived and accordingly some genes are linked to tumor activity or genes that are linked to species-specific characteristics and are dissimilar to the human situation when looking at animal models).

Gene expression profiling can be used to characterize such assays. For example, when one is interested in the modulation of a specific receptor, it is important to know whether there are other receptors of the same family being expressed in the cellular system. Another aspect are the presence or absence of key components downstream of the molecule that is to be modulated. In case the modulation cannot be measured directly, it might be possible to obtain read-outs based on the effects of the modulation of other molecules downstream.

7.3.5   Characterization of a disease model

Currently the typical approach for identifying a molecule that can be developed into a new medicine requires the identification of a molecular target. While there are different elements that contribute to the qualification of a given protein as a suitable target to modulate a given disease (such as literature knowledge), one element that is typically required is the availability of a model. One such model can be, e.g., a knock-out mouse of the candidate target gene. In case the gene knock-out is not lethal for the animal, a large battery of experiments is often done to characterize the phenotype of the animal as good as possible, as the effect of the gene knock-out is rarely visible to the naked eye. Applying gene expression profiling in an experiment that compares knock-out animals vs. their wild type littermates is an approach to understand more about the long-term effects of the gene knock-out on the animal. This typically identifies the different compensation mechanisms which allow the animal to deal with the chronic absence of the gene-encoded protein.

Similarly such experiments can be done for genetically modified mice that overexpress a certain gene. In a study by Peeters et al., mice chronically overexpressing the gene encoding for the corticotrophin releasing factor (CRF) were studied to understand more about the differences between the genetically modified animals and the wild type controls[39]. CRF is an important factor in mediating the response to stress. Alterations in the CRF system have been linked to depression. Animals overexpressing CRF were considered as a potential model of one aspect of depression as patients with depression are among others characterized by an increased level of CRF in the cerebrospinal fluid and by an increase in neurons secreting CRF. When looking at the phenotype of the genetically altered mice, they can be characterized by a number of typical symptoms that are associated with the life-long exposure of the animals to elevated levels of CRF:

1. Increased heart rate and body temperature.

2. Reduced activity when the animals are exposed to a brightly illuminated open area as measured by the behavioral test called "open field test." This test models how much a mouse reacts to such an anxiogenic stimulus by measuring its exploratory behaviour, e.g., total distance, speed, time spent in border areas vs. open areas in the middle of the field.

3. Reduced startle reactivity and prepulse inhibition. This test is often performed to assess whether the animal can filter out unnecessary sensory signals or whether the information processing is impaired. In essence a softer stimulus (prepuls) is given prior to a stronger stimulus (puls) which results in a weakened startle response. Normal animals will have a smaller startle response after such a prepulse as compared to animals that did not receive a weaker pulse initially.

The gene expression study demonstrated the compensatory efforts of the animals that had to cope with life-long increased amounts of CRF. For example, 11β-hydroxysteroid dehydrogenase is a gene converting inactive forms of CRF into the active form. This gene was clearly downregulated in the animals to reduce the amount of CRF to some extent. Similarly, Fkbp5 is a gene encoding a potent inhibitor of glucocorticoid receptor binding. The clear induction of these genes is another mechanism by which the animal tries to counteract the effects of elevated levels of CRF by inhibiting the receptors that in turn would lead to more secretion of CRF.

However the study also showed that these animals are rather a model for chronic stress. A compound intended to have a positive effect on the CRF levels in the over-expressing mice was only able to prevent an increase in CRF due to stress caused by the animal handling. It was not possible to bring the elevated levels of CRF in the genetically altered animals back to a level similar to the wild type animals. In other words, the gene expression study revealed clear limitations of this animal model as a model for the CRF-related aspect of depression.

Another area that is getting increasing attention in pharmaceutical research is the use of phenotypic screens. Typically these screens will involve a cell line assay which exhibits a desired phenotype. For example, when searching for a compound that could potentially inhibit the formation of metastases, such an assay could be a migration assay. Here, cells would be cultivated in such a way that one could observe migration of cells under the microscope. A screen using thousands of compounds would search for compounds that inhibit migration. An essential difference of such a phenotypic screen is the intentional lack of knowledge about the underlying molecular mechanisms. One is not interested in modulating the activity of one or few target proteins but rather in changing a certain characteristic that is thought to be of relevance for a given disease. Again, gene expression profiling studies can be used to characterize these phenotypic assays. Knowing potential candidate genes that are involved in the phenotype narrow down the candidate genes and proteins that can be investigated once a potent compound is identified. This is crucial information as an essential requirement for registering a new medicine is knowledge about the mechanism of action of the drug.

7.3.6 Compound profiling

Compound profiling experiments are commonly applied in pharmaceutical research. They can be very valuable to discover differences in mechanism of action between various compounds. To illustrate this type of research question, we have used the data of Dillman et al.[10] as an example. Dillman et al. compared three treatment compounds that have been shown to limit the damage induced by the alkylating agent sulfur mustard (SM) in a mouse ear model: dimercaprol (DM), octyl homovanillamide (OHV) and indomethacin (IM). With microarrays they determined gene expression profiles of biopsies taken from mouse ears after exposure to SM in the presence or absence of either one of the three treatment compounds.

When designing compound profiling experiments one has to make a number of experimental design decisions. Regarding the choice of compound concentration one can either fix one constant concentration for all compounds or choose the concentration based on an independent readout such the IC50 concentrations derived from an apoptosis assay.

Furthermore, the choice of cell line should also have a link to the target disease. And to avoid spending time unnecessarily during data analysis when

EC50 (effect concentration 50%) refers to a measure of clinical efficacy of a drug. It defines the concentration which exhibits 50% of the maximum effect. The term is used for both inhibitory as well as stimulatory effects.

ED50 (effective dose 50%) defines the median effective dose at which 50% of individuals show a desired effect.

IC50 (inhibitory concentration 50%) is the concentration at which an inhibitory compound produces 50% inhibition of, e.g., the activity of an enzyme or of growth.

LC50 (lethal concentration 50%) is the concentration of a cytotoxic drug where 50% of cells die.

LD50 (lethal dose 50%) is a measure of how much of a substance constitutes a lethal dose in half the test population.

BioBox 7.2: EC50, ED50, IC50, LC50 and LD50

trying to make sense of a result that does not seem to make sense, it can pay off to have performed an independent control experiment in parallel with exactly the same compound solutions. This way one can ensure that all compounds were active in their respective solutions.

7.3.7 Dose-response treatment

Studying the molecular downstream effects of modulating a receptor with a compound requires an optimal dose that has the desired effect on the receptor while not leading to toxic side effects. As different compounds also exhibit different active doses (see also BioBox 7.2), it is often difficult to choose the right concentration. If affordable and technically possible, the best solution is to use multiple doses. This can be even more helpful when one expects a sigmoidal curve whereby lower concentrations have almost no effect and concentrations above a certain level plateau the response from the receptor. Using this assumption one can specifically search for genes that show such sigmoidal induction or repression of gene expression by attempting to fit a log-logistic model on the expression levels of each gene across a number of different concentrations of an administered compound. This approach will focus on genes that show a true dose response effect, as the number of genes that will have such a characteristic behaviour by chance is very small.

For example, in one experiment we looked at alterations downstream of the dopamine D2 receptor. The pharmacological activity of antipsychotics is often based on blocking this receptor. However, it is also this blocking that can cause side effects such as extrapyramidal symptoms (EPS) and hyper- prolactinemia (elevated prolactin level in serum) in humans. Two measures that can be used to estimate compound activity are antagonism of apomorphine-induced behaviour and receptor occupancy in the brain.

An animal model that can be used to study human EPS is the sensitivity to catalepsy in rats. We administered a number of different antipsychotics at six different concentrations. These concentrations were based on the ED50 value as determined in an independent apomorphine test. Apomorphine is a direct dopamine receptor stimulant which mimics the agonistic action of dopamine on the D2 receptor. The resulting behavioral effects can be antagonized by blockade of central D2 receptors. The efficiency of this antagonism can be a measure for the potency of a compound.

After an accute treatment of 1h, we isolated a specific, relevant brain region (the striatum) and performed gene expression profiling on all samples. During the data analysis it was attempted to fit sigmoidal dose response curve onto all genes. Using this approach, we were able to identify less than 100 genes with a clearly sigmoidal curve. Figure 7.2 illustrates the sigmoidal dose-response behavior of the transcription factor Fos for some of the antipsychotics. Fos is known to be a downstream target of dopamine receptors[241].

This example demonstrates how the biological question (which genes are affected downstream of the dopamine D2 receptor?) could be used to define an analysis strategy (make use of a sigmoidal dose-response curve) that in turn influenced the experimenal design. It identified the few genes on which antipsychotics have a clear dose response effect.

7.4 Drug development

7.4.1 Biomarkers

Scientists are interested in developing reliable sets of measurable characteristics (biomarkers) that correlate with specific clinical outcomes such as physiological, pharmacological (compound efficacy), toxicological or disease processes. The identification of such biomarkers offers the possibility that they could be translated into diagnostic tests. These tests in turn will be used as decision-making tools for doctors on, e.g., the most appropriate treatment for a given disease. Thus, the term biomarker is used for various purposes (see BioBox 7.3). One group of biomarkers is intended to provide the practitioner with a basis for decisions in the clinic like:

1. Who should be treated?

2. How should a patient be treated?

3. With which medication should a patient be treated?

Markers and signatures. Besides the general definition of a biomarker or a signature, there are a number of other names being used for different contexts.

Disease markers potentially help us understand more about the pathology of a disease. They are primarily intended for diagnostic and prognostic purposes.

Prognostic markers characteristics of a patient or tumour at the time of diagnosis that can be used to estimate the chance of the disease recurring in the absence of therapy.

Efficacy markers indicate the ability of a drug to induce a biological response in its molecular target.

Patient response markers aim to predict whether a patient will respond to a given treatment or not.

PD marker refers to pharmacodynamic (PD) marker. This is a molecular drug activity marker that can be measured in patients receiving a drug. The marker should be a direct measure of modulation of the drug target and be able to show quantitative changes in response to dose.

PK marker stands for pharmacokinetic (PK) marker. This drug activity marker is used to indicate systemic exposure to an investigational drug.

Response markers Characteristics of a patient that are associated with the response to a particular therapy.

Surrogate markers/surrogate endpoints are biological measurements that are intended to correlate with a true clinical endpoint. These markers are defined for mainly two reasons: (1) obtaining the true endpoint is undesireable or unethical (e.g., death), (2) the true endpoint is impossible to measure (e.g., very low frequency of occurance). Surrogate markers are typically used as primary measures of the effectiveness of drugs.

Toxicity biomarkers are intended to aid in understanding and developing drug dosage parameters to avoid toxic effects in the patient.

BioBox 7.3: Markers and signatures

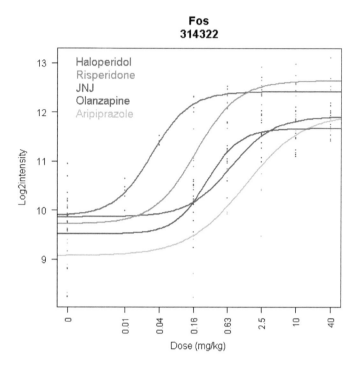

FIGURE 7.2: Sigmoidal dose response curves of Fos gene expression changes in rat striatum after treatment with different antipsychotics. The different colors represent the different antipsychotics.

Another group of markers is used by pharmaceutical companies for internal use to prioritize compounds. PK/PD markers are examples of such biomarkers. The level of validation and qualification of new biomarkers needs to be adjusted according to the field in which they are to be applied: will the marker be used for internal decision-making for diagnostic purposes? On the other hand, disease specific biomarkers will need to fulfill different requirements than safety related biomarkers.

Currently biomarker validation typically involves testing a small number of candidate biomarkers in a large population of samples. However, among other technologies, gene expression studies using microarrays are increasingly used for this purpose as well. As they measure thousands of genes simultaneously, they offer the possibility to discover complex, multivariate markers. Such multi-gene signatures are thought to better reflect and capture complex diseases such as cancer. However, the robustness and reliability of array-based signatures, as well as the clinical utility, has to be shown. As assay conditions are likely to be variable (e.g., deviations from standard operating procedures

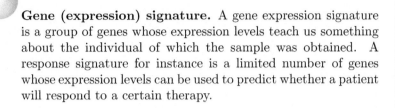

Gene (expression) signature. A gene expression signature is a group of genes whose expression levels teach us something about the individual of which the sample was obtained. A response signature for instance is a limited number of genes whose expression levels can be used to predict whether a patient will respond to a certain therapy.

BioBox 7.4: Gene (expression) signature

or differences in patient populations), it is important to have robust prediction models that show discriminative power between prediction groups, superiority over existing tests and improved clinical outcome.

Any given marker is typically judged by a set of four analytical performance parameters: test accuracy (how close is the measurement to the true value?) and precision (how much variability is associated with a given measurement technique?) as well as analytical sensitivity (are all patients that are truly suffering from a disease actually identified by the technique?) and specificity (are patients identified as being positive for the test while in fact they are not affected?).

7.4.2 Response signatures

A major challenge in the field of oncology are the various levels of response across the whole patient population when administering a chemotherapeutic. As especially oncological agents can have side activities that affect the patient's quality of life, it has become clear that future targeted agents ought to be given only to patients who will benefit from the treatment. This goal requires the availability of markers predicting the sensitivity of a patient to a given therapy. A number of recent studies have demonstrated the successful application of gene expression studies for the identification of signatures that predict chemotherapeutic response (e.g., [242]). The ability to identify response signatures is key to move cancer treatment to the next level whereby clinical outcome is improved and unnecessary treatment of unresponsive patients is avoided (see also 7.4).

The experimental procedure usually starts with the generation of whole genome expression profiles of untreated cell lines and/or primary tumors. Using various statistical approaches, one identifies a group of genes that classifies cell lines or tumors on a molecular level into responsive and non-responsive subtypes.

The problem with this approach is that the signature needs to be truly predictive as it is used at a stage when the patient has not yet been treated with the compound. In other words, only after people were treated with the chemotherapeutic and the responsiveness is obtained, one will know whether the signature used for patient selection was indeed predictive of the response.

One strategy applied successfully is the use of tumor cell lines which are thought to still reflect key characteristics of the tumor from which they were derived (e.g., [243]). Here one looks at the correlation of gene expression data of various cell lines grown under normal conditions without compound treatment with the responsiveness of the same cell lines to a compound measured in a separate *in vitro* experiment. Any identified signature is subsequently validated using a new set of cell lines of which the responsiveness has been predetermined.

Even though microarrays have demonstrated their potential as a discovery tool for response signatures, one has to keep in mind that they are currently not being considered as a clinical test. Therefore the validation of any potential microarray-based gene signature needs to be done with a technology that is suitable for the clinic, such as real-time quantitative PCR.

7.4.3   Toxigenomics

The name of this scientific discipline reflects the attempts to make use of a combination of toxicology with genomics technologies such as gene expression profiling and bioinformatics. By using whole genome approaches, possible adverse effects of compounds are studied and predictions are made regarding the toxic potential of novel molecules. Typically, samples from liver, kidney and blood are collected and the gene expression profiles are determined.

In pharmaceutical research, the aim is to have an accurate toxicological characterization of compounds early. The compounds with the best toxicological profiles are to be selected as starting points for chemical modification and further development. This way information can be obtained on drug safety at a stage when only small amounts of the compound have been synthesized. Identifying toxicity prior to a scale-up in compound synthesis is very beneficial for a pharmaceutical company as the costs associated with the scale-up are much larger as compared to the price that needs to be paid for the toxicogenomic study.

A major promise of using gene expression profiling for toxicity studies is the potential to identify gene activity alterations already at a stage when the effects can be seen on the mRNA level but have not manifested themselves to an extent that would be identifiable by classical toxicological investigations such as histopathology.

While there has been significant progress on the technological side, obtaining meaningful information from toxicogenomic studies remains challenging. Current publications primarily describe alterations in, e.g., transcript levels but have so far provided only limited insights into the mechanisms of toxicity. Furthermore, in contrast to early expectations, toxicogenomics has not proven to be more sensitive in identifying toxic effects. However, it does seem to keep the promise of identifying toxic effects earlier than classical toxicological investigations.

Examples

Predictive markers for toxic effects Toxicogenomic studies can be used to develop markers that predict whether a certain compound will be toxic to humans. In other words, the main goal of such a marker is to identify the toxicity of a compound with high accuracy and specificity.

The starting point for achieving this goal is the construction of a database with gene expression profiles from studies on hundreds of compounds with a known toxicological profile, e.g., various carcinogens. Based on this data a marker is developed to predict the carcinogenic potential of a new compound with unknown toxicology properties. Alternatives to identifying a marker are strategies by which compounds with an unknown toxicological profile are clustered together with all compounds in the database. The group of compounds with which the compound most closely co-clusters and of which the toxicity is known is thought to predict the same type of toxicity for the new chemical entity.

Backup strategy Another example is the use of gene expression profiling to optimize the backup strategy for a compound. While the lead compound is pushed forward, gene expression profiling might indicate an influence on the activity of nuclear receptors such as PPARα. Assessing the set of backup compounds for their impact on these receptors and prioritizing those that do not have the same effect as the lead compound can turn out to be beneficial when the lead compound does indeed show toxicity linked to an effect on PPARα.

A typical design involves an animal treatment with a compound under controlled conditions, e.g., the dosing of male rats for three days using two different doses per compound. The compound doses are chosen based on two characteristics: one dose is expected to be pharmacologically efficacious and the other dose is as high as possible while avoiding a lethal dose. These two extremes are intended to cover the range of effects that one can obtain when treating with the tested compound.

Mechanism of action related toxicity The aim of studies done in this context is to enhance the understanding of the mode of action of a known toxicant. This is in contrast to the development of a toxicity marker. Understanding how different toxicities arise when studying a broad spectrum of reference toxic agents is the basis for making better hypothesis about what kind of toxicities could be induced by a compound with an unknown toxicological profile. Furthermore, this might also assist in estimating the relevance of potential toxic effects to humans.

InnoMed PredTox   The InnoMed PredTox consortium was initiated by the European Federation of Pharmaceutical Industries and Associations (EF-PIA). This joint industry and European Commission collaboration is looking into a limited number of compounds that failed during pre-clinical development due to issues that came apparent in conventional toxicity tests. Participants from Pharma and Biotech companies as well as from universities and technology providers seek to identify the causes for these failures to establish approaches that can be applied to prevent future failures of compounds at a late stage when substantial resources have already been invested. Furthermore, the consortium also attempts to make recommendations on the most ideal combination of methods and technologies for future drug development projects. As the experiments involve not only data from gene expression profiling but also results from proteomics, metabonomics and conventional end-points, attempts are made to integrate the analysis and storage across the different data types[244].

7.5   Clinical trials

7.5.1   Efficacy markers

In essence an efficacy marker (also referred to as "response marker") is intended to predict the outcome of a clinical trial early on in the study. Popular examples of such a marker are: skin rash for EGFR inhibitors such as gefitinib (Iressa®[3]) or viral load when studying the effects of drugs against viruses such as HIV or Hepatitis C.

In the context of this book, we are of course interested in searching for markers based on gene expression alterations. The ideal marker should indicate a change in expression of selected marker genes measured in the blood of patients after a treatment has been applied.

7.5.2   Signatures for outcome prognosis

While predicting the outcome of a treatment is especially important in life-threatening diseases like cancer, the actual prediction remains very difficult providing only rough indications about outcome or recurrence. Clinicians typically make use of information like tumor size and spread, tumor subtypes, etc. for assessing the aggressiveness of a tumor. In this exercise, the disease stage (histopathological diagnosis) often turns out to be most relevant for therapeutic decisions prognostic estimation.

---

[3]Registered trademark of AstraZeneca, London, England.

Dataset OP is one example where scientists have used gene expression profiling to come up with a marker that will better predict outcome for patients with advanced bladder cancer. In the end the scientists were indeed able to show that a combination of two marker genes (Emmprin and Survivin) had a strong prognostic value for both response to the treatment as well as survival time[14].

Similarly, a study by Nutt et al.[245] looked into better classifying malignant gliomas using gene expression data. Their efforts were motivated by the fact that this type of cancer is diagnostically challenging when based on histologial features. While the sample size was rather small (n = 50), the model they built was able to define two tumor subclasses that were significantly associated with survival outcome (P = 0.05). In contrast, when looking at tumors that were classified into two groups according to pathology (glioblastomas vs. anaplastic oligodendrogliomas), the predicted survival of patients was not significantly different (P = 0.19). This example resulted in the development of a prediction model that could provide clinicians with a more objective and more accurate classification approach as compared to the traditional histopathological classification. In other words, the gene-expression derived model divides the gliomas in two classes that better correlate with clinical outcome than standard pathology.

While the example by Nutt et al. described the results from an initial finding, there are already a number of gene expression derived markers for outcome prognosis in clinical trials to confirm their prognostic value (e.g., gene signatures for breast cancer[246],[247]). This highlights the relevance of using gene expression studies to discover such markers and stresses the potential medical relevance of such molecular markers as compared to classical histopathological diagnosis.

Chapter 8

Using R and Bioconductor

This book has been made using open-source software products. We used R 2.8.2 [3] and Eclipse for data analysis, Inkscape 0.46 [248] and The Gimp 2.6 [249] for making the graphics and LyX 1.5.6 and LaTeX for text editing. The Web links to the sites of these software tools can be found on the website of this book. This chapter solely focuses on the software used for the data analysis, namely R.

R is a language and environment for statistical computing and graphics. It is an integrated suite of software facilities for data manipulation, calculation and graphical display. R is available as free software under the terms of the Free Software Foundation's GNU General Public License in source code form. R is nowadays widely used because of its many strengths and advantages, most of which are listed below.

Philosophy. The core developers do not consider R as just an analysis system, but prefer to think of it as an environment within which statistical techniques are implemented. The term "environment" is intended to characterize R as a fully planned and coherent system, rather than an accumulating bunch of very specific and inflexible tools.

Possibilities. The range of possibilities with R (and Bioconductor) packages is dazzling. Probably every analysis or graph can be produced in R. Even the most cutting-edge techniques are often available since most statisticians and bioinformaticians provide a package together with the publication of their approach. Because it is not a menu-driven software, the user must however be familiar with programming in R.

Language. The R language is a well-developed, simple and effective programming language with many coding tools and many input and output facilities. It is computationally very efficient and allows to make direct use of C(++) or Fortran code for computationally intensive tasks.

R can be extended easily via *packages*, which are standardized extension units to the R core packages. There are currently more than 1,600 packages available through the CRAN network of Internet sites covering a very wide range of traditional and modern statistics. Packages can easily be installed in R by typing the following:

```
> install.packages(namePackage)
```

Graphics. R can easily produce well-designed publication-quality plots with fancy colors and symbols. The default settings for the simple design choices of the graphics have been chosen with great care, but the user retains full control. The graphics can be displayed on-screen or can be exported as a file in all common formats (pdf, png, ...).

Documentation. R has a comprehensive documentation, both in hardcopy and on-line in a number of formats including html and pdf. Every function or package is documented. To display the help file of a respective function within R (for example the function `image`), one only needs to type it preceded by a "?". The *help file* of a function typically contains a lot; it describes what the function does, how it should be used, which arguments can be used in the function, some details, links to other relevant functions, references and finally some examples.

```
> ?image
```

Every help file typically contains several examples to get the user easily started. The code of these examples – and consequently its resulting output – can easily be invoked within R directly with underlying code (here again for the function `image`).

```
> example(image)
```

Besides the help files many packages, especially Bioconductor packages, are accompanied by *vignettes.* R help files are very useful, but they often do not demonstrate how multiple functions can be used and combined. Package vignettes show all the relevant steps of a particular analysis and, along the way, explain the functionality of most functions, and are therefore practically more useful compared to help files.

8.1   R and Bioconductor

Bioconductor[250] is a set of packages for bioinformatics, and more particularly for the analysis and comprehension of genomics data. It is an open source software project, just like the software R in which it is rooted. Because of the large enthusiastic community working in and with it, it is by far the most comprehensive platform for gene expression analysis.

Most Bioconductor components are distributed as R packages, which are add-on modules for R. These packages can easily be installed in R by typing the following:

```
> source("http://bioconductor.org/biocLite.R")
> biocLite()              # installs standard BioC packages
> biocLite(namePackage)   # installs additional BioC package
```

8.2 R and Sweave

Another important reason for our fondness of R is the availability of the Sweave functionality[251]. Sweave is a function in R that enables integration of R code into LATEX documents. This makes it possible to jointly document the R analysis code and write the report of the analysis results and conclusions. The purpose of Sweave is to create reports that are reproducible, transparent and dynamic.

Analyzing data and writing reports have typically been two separate stages. First, data is analyzed using some statistical software based on data files and code files (the latter if the statistical software was not a "point&click" software). Only after the results have been obtained, they are used as a basis for a written report, resulting in a separate file containing the report (a Word-doc or a pdf-file). Sweave allows to embed the analysis into the report, so that there is only a report and some data files.

Sweave allows to build a dynamic document as an ordered composition of code chunks and text chunks that describe and discuss a problem and its solution[252]. Code chunks consist of R commands that produce the appropriate output within the paper. The text chunks contain the text and titles that make up the actual report. Typically, this text will describe the problem, the results and their interpretation.

There are basically three main advantages of the type of reproducible reports built by Sweave (see Section 6.4 for the context of the problems).

1. The results of a given analysis can be evaluated by other researchers in the public domain. It has been remarked several times that the field of microarray data analysis suffers from irreproducible findings, even the results of published studies[183],[184]. The use of Sweave will solve this issue, thereby offering a quality check tool. If the quality of scientific research can be evaluated more thoroughly, this may enhance scientific evolution as false positive findings due to flawed analysis strategies will be unmasked relatively easy and early.

2. There is a high level of documentation of the code. As the R code is embedded in the report, it is easy to put the code and its resulting output in the context of the study. This makes the analysis much more transparent, thereby facilitating for instance the exchange of code between researchers.

3. An important practical advantage is the ease by which a full report, starting from data, can be reused. Both code and text can be adopted to generate reports of studies with similar designs and data structures. This dynamic nature of analysis and report writing enables one to work more efficiently and to save a lot of time. To highlight how Sweave might improve your daily work, Scott[253] presented three awkward situations which would have been avoided by using Sweave:

   (a) You sit down to finish writing your manuscript. You realize that you need to clarify one result by running an additional analysis. First, you re-run the primary analysis, but the primary results don't match what you have in your paper.

   (b) When you go to your project folder to run the additional analysis, you find multiple data files, multiple analysis files, and multiple results files. You can't remember which ones are pertinent.

   (c) You've just spent the week running your analysis and creating a results report (including tables and graphs) to present to your collaborators. You then receive an email from your client asking you to regenerate the report based on a subset of the original dataset and including an additional set of analyses; she would like it by tomorrow's meeting.

Code is available on-line. Every analysis or data graphic in this book has been made using R. We have deliberately not shown or discussed this code in the book to keep the focus on the content. All our code is however made available on-line on the book's website `http://expressionstudies.com`. This example code might be helpful when exploring certain methods described in the book or when trying to apply them on one's own data.

To allow people to reproduce the graphs, we have carefully chosen to only use publicly available datasets.

8.3 R and Eclipse

We currently use Eclipse with the StatET plugin to work with R because of its user-friendly interface. Eclipse is standard open-source software for Java development that allows easy access to code management systems like bug and feature tracking systems. The StatET plugin neatly integrates R consoles and

R scripts within Eclipse. It also allows error detection and code assistance as it recognizes the R syntax and keywords.

 Bite the bullet. Both learning R as well as learning LaTeX can be quite demanding. However, as both work very well together and create enormous potential for making data analysis very efficient, we highly recommend learning these languages.

8.4 Automated array analysis

We recommend using packages of the original authors. Although their adoption may sometimes take a little time, it is worth the effort. An additional advantage is that these packages are often well documented with help files and vignettes. Particularly vignettes will contain valuable suggestions and warnings against potential misuses. Furthermore, the functions within these packages are generally carefully constructed by avoiding redundant functions to keep the overall process user-friendly. Functions that may only be sometimes useful will however remain, as one wants to keep methodology flexible and generalizable.

For our own work, we therefore have packed frequently used chunks of code into single functions. This facilitated the re-use and exchange of code and made the entire analysis process more surveyable. We have bundled these functions into the package a4, standing for "Automated Affymetrix Array Analysis." The package a4 is a wrapper of many functions from different packages, and has consequently many dependent packages. As the original packages are used, the analyses are still based on the original code from the authors. Only the steps that typically remain unchanged over different studies are hidden and are running in the background. The package also contains a Sweave document that can serve as a template for reports for standard analyses such as comparisons between two groups.

A detailed description can be found on the book's website, but the main building blocks are shown below.

8.4.1 Load packages

The code below loads the package a4. See guidelines on the website for other details including how to install R and the relevant packages and how to import the data. The data can come in many forms like text files, expression sets and CEL files. The latter, which are the raw Affymetrix data (see Section 4.1.1), are incorporated into a so-called AffyBatch object.

```
> library(a4)
```

8.4.2   Gene filtering

The code below filters the AffyBatch object using I/NI calls[97] (see Section 5.3), and summarizes the probe intensities into gene expression values using FARMS[75] (see Section 4.1.5). This step only works for Affymetrix data.

```
> exprSet <- iniCalls(AffyBatchObject)
```

8.4.3   Unsupervised exploration

The code below shows a spectral map[139] of the filtered expression data exprSet (the two first dimensions) and colors the samples according to the variable Group (see Section 5.4.3).

```
> spectralMap(exprSet, "Group")
```

8.4.4   Testing for differential expression

The code below invokes a test for differential expression between the two classes of the variable Group using an ordinary t-test (see Section 5.5.2.2.1. Next, the corresponding volcano plot (see Section 6.1.5.1) is shown and the 20 most significant genes are listed.

```
> resultTtest <- tTest(exprSet, "Group")
> volcanoPlot(resultTtest)
> topTable(resultTtest, n = 20)
```

The code below invokes a test for differential expression between the two classes of the variable Group using a modified t-test (Limma[156], see Section 5.5.2.2.3), shows the corresponding volcano plot and lists the 20 most significant genes.

```
> resultLimma <- limma2Groups(exprSet, "Group")
> volcanoPlot(resultLimma)
> topTable(resultLimma, n = 20)
```

Finally, show the expression levels of the top-ranked gene by Limma for all samples, colored by the variable Group, using an intensity plot (see Section 6.1.2).

```
> plotProfile(exprSet, "Group",
+         feature = rownames(resultLimma)[1])
```

8.4.5 Supervised classification

As described in Section 5.6, there are many different ways to classify samples based on gene expression and to search for predictive sets of genes. Therefore, we suggest to apply several different algorithms. Here, we use the four complementary and valuable classification methods which were recommended in Section 5.6.4.9.

The code below invokes forward filtering in combination with various classifiers (like DLDA, SVM, random forest, etc.) in a nested loop cross-validation. The selection is here based on Limma[156], so that a univariate and independent feature selection is applied (see Section 5.6.3 on page 204 for more details).

```
> nlcvLimma <- nlcv(object = exprSet,
+               groups = "Group",
+               nRuns = 20,
+               fsMethod = "limma")
> mcrLimma <- mcrPlot(nlcvLimma)
> scoresPlot(nlcvLimma, optimal(mcrLimma))
> topTable(nlcvLimma, n = 20)
```

The code below invokes a PAM classifier[185] containing a univariate and dependent feature selection method (see Section 5.6.3 for the output generated by this code and Section 5.6.4.8 for more details on PAM).

```
> resultPam <- pamClass(exprSet, "Group")
> plot(resultPam)
> topTable(resultPam, n = 20)
```

The code below invokes random forest with variable importance filtering[182], applying multivariate and dependent feature selection (see Section 5.6.3 for the output generated by this code and Section 5.6.4.7 for more details on random forest).

```
> resultRF <- rfClass(exprSet, "Group")
> plot(resultRF)
> topTable(resultRF, n = 20)
```

Finally, the code below invokes LASSO[185], a classifier with a multivariate and dependent feature selection approach (see Section 5.6.4.3).

```
> resultLasso <- lassoClass(exprSet, "Group")
> plot(resultLasso)
> topTable(resultLasso, n = 20)
```

8.5 Other software for microarray analysis

Besides R, there are of course numerous other commercial, freeware and open-source programs available for microarray data analysis. Discussing all of these would fill a book on its own. However, here is a short overview of programs that are often cited in publications. The list is not complete and certainly does not reflect a judgement about the quality of the software that is included or not included.

dChip The dChip software from the labs of Cheng Li and Wing Wong aims to provide a tool focused on the analysis and visualization of gene expression and SNP microarray data. It is written for the Microsoft Windows platform and focuses on the Affymetrix platform. However, other platform data is supported as well. Some of the major analytic capabilities of dChip software are normalization (invariant set-normalization based on a subset of PM probes that show a small within-subset rank difference between arrays), model-based expression intensities (using a PM-only model or a PM-MM model), gene filtering based on genes showing large variation across samples or being present in most samples, hierarchical clustering of genes or samples, finding genes with high/low fold change/difference, and some set operations, finding genes with a similar time-course expression pattern to a given gene (correlation) and finding significant GeneOntology, ProteinDomain, or Cytoband similarities in a given gene list. The software can also be used for copy number analysis of SNP data, visualizing SNP data along the chromosomes and performing linkage analysis.

Cluster and TreeView Both software packages and a number of other tools for cluster analysis and visualization as well as microarray image analysis, gene expression and sequence analysis are made available through Mike Eisen's lab at Stanford. Cluster performs a number of different cluster analysis, including hierarchical clustering, k-means clustering, and self-organizing map construction. The dendrogram results for hierarchical clustering can be viewed and further analysed in the complementary software TreeView.

GeneCluster GeneCluster has been made available through the Whitehead Institute at MIT. It performs supervised classification, gene selection and permutation tests. There is a module for batch self-organizing map clustering and supervised models can be built and tested by applying algorithms such as weighted voting (WV) and k-nearest neighbors (KNN). It implements the analysis approach that was used in the publication by Tamayo et al.[118]. As the support of the program has been stopped, people interested in using the algorithms are asked to use the software GenePattern which has further analysis and workflow capabilities.

GeneSpring GeneSpring is commercial microarray data analysis program originally developed by Silicon Genetics. Since the acquisition of Silicon Genetics by Agilent, the software is now called "GeneSpring GX." Among many features, it performs hierarchical clustering, gene list construction by similar expression pattern, functional classification, regulatory sequence searching, and some pathway construction. The marketing of the product highlights the focus of the software on the needs of biologists.

Pathway studio This commercial software developed by Ariadne Genomics has among many other features the ability to perform gene expression data analysis with a special focus on the biological context (e.g., pathways, gene regulatory networks, biological processes, etc.). Gene expression data can be covisualized with known pathways and new pathways can be constructed based on literature data.

Chapter 9

Future perspectives

9.1 Co-analyzing different data types

With the establishment of more and more molecular profiling technologies, it becomes increasingly interesting to combine profiling data obtained from the same sample. These include data from different microarray technology based experiments (e.g., gene expression, copy number, DNA methylation; see also BioBox 9.1), but also data from mass spectrometry-based technologies for metabolite profiling, proteome analysis, etc. or next-generation sequencing.

Current strategies typically involve a separate analysis and interpretation of the different data. The results are subsequently visualized across the different data types. By looking at "interesting" data identified in one data type together with the matched data from the other profiling experiments, researchers find data that agree with one another. It is hoped that the combination of relevant markers originating from different molecular characteristics present in a cell (e.g., identifying a set of 5 over-expressed transcripts in a region with an increase in DNA copy number that results in higher amounts of protein) will be more conclusive than single or even a set of measurements from a single molecular type (e.g., mRNA levels).

For example, in a study by Garraway et al.[254] the researchers looked into skin cancer caused by malignant melanocytes (pigment producing skin cells). It was concluded that those skin cancers that had reduced activity of a certain protein (MITF or microphtalmia-associated transcription factor) in the cancer cells were more effectively killed by chemotherapeutic drugs than others. To reach this conclusion, the authors combined DNA copy number results from SNP microarrays with gene expression microarrays. After clustering the copy number data, they identified an amplified genomic region between 3p13 and 3p14 (a location on chromosome 3). Once this region was discovered, the subsequent gene expression analysis focused on determining whether highly expressed genes in this region might be identified as a putative oncogene targeted by the amplification. In other words, the two different data were analyzed sequentially and not together in one combined analysis.

Future research in the field of data analysis both in the bioinformatics as well as in the biostatistics world should therefore develop strategies how these different data types and data formats could be analyzed together rather than

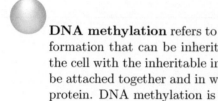

DNA methylation refers to a modification of the genomic information that can be inherited. The DNA sequence provides the cell with the inheritable information which peptides should be attached together and in what order to produce a functional protein. DNA methylation is an inheritable modification (also referred to as an "epigenetic modification") that carries information about gene activity. In general, if certain parts of the DNA for a given gene are methylated, the gene will be inactivated. In other words, adding or removing the methylation of certain DNA regions will change the activity of the gene but not the information that is encoded in that region. Changes in DNA methylation is a characteristic that is often seen in cancers, where an increase in methylation results in the inactivation ("gene silencing") of genes that otherwise could have stopped the growth of the cancerous cell.

BioBox 9.1: DNA methylation

sequentially or just co-visualized. We need to develop algorithms that can standardize data of different characteristics and process and analyze them jointly. This would be especially appealing in the context of reproducible research as strategies based on co-visualization would most likely yield different results for different researchers.

9.2   The microarrays of the future

It is currently not clear how much more the feature size can be reduced. Looking at the evolution of Affymetrix arrays, the feature size of standard, commercially available arrays has evolved from 50 micron to features of 4 micron with the current SNP 6.0 arrays. As the feature size decreases, the amount of data generated per sample increases exponentially. Of course this has an impact on computer hardware needs and processing times.

While feature reduction is a way of increasing the number of probes per array, another important area of research goes into improvements that would result in an increase in signal strength. Besides improving the sensitivity of the technology for low abundant transcripts and transcripts which have poor hybridization characteristics due to low AT-content in the sequence, it would again allow for a further decrease in feature size. As features become smaller, the overall signal decreases. This is especially an issue for low abundant transcripts. One possibility is an increase in the number of probes per feature by switching to a three-dimensional structure as opposed to a two-dimensional synthesis on the surface of the waver[255]. Such three-dimensional features can be made of, e.g., sugar hydrogels, nitrocellulose membranes, porous silica films or layers of long-chain, hydrophilic polymers which have already been successfully commercialized in the past by, e.g., Motorola's CodeLink®[1] arrays.

Another field of research at Affymetrix has been new chemistry techniques that not only result in higher yields but also allow for the synthesis of oligonucleotides of up to 75 bases. There have been plans made public to release a set of three arrays with 30 million 50-mer probes for ultra-high resolution copy number analysis. Each array is designed to contain 10 million probes covering only the non-repetitive areas of the human genome.

Using enzymatic reactions on the array can be another future direction, especially in the field of sequencing products whereby, e.g., labeling reactions carried out by polymerases and ligases are done inside the cartridge. With the recent acquisition of True Materials, Affymetrix will also investigate technologies that are similar to other bead-based platforms such as the one marketed by Affymetrix's major competitor Illumina. It is anticipated that future arrays (also referred to as "liquid arrays") that would make use of True Material's barcoding technology would be combined with applications in the low to mid-multiplex range (100 to 10,000 markers). Such arrays are thought to be used for applications in the field of gene expression, SNP genotyping, miRNA profiling, protein profiling, methylation profiling or food/environmental testing and molecular diagnostics.

Furthermore, with 2 FDA approvals already received for Affymetrix technology (one for the instrumentation platform 3000Dx and one for a test that identifies the tissue origin of a tumor sample by looking at the gene expression profile of the tumor sample), it is quite likely that future arrays and applications will accommodate diagnostics applications.

Lastly, with the recent announcement of the GeneTitan®[2] next-generation array platform, Affymetrix is moving into the field of high-throughput sample processing. Instead of single cartridges per sample, this platform makes use of so-called "PEG arrays" which are small arrays (approximately one fourth of the size of current arrays) located at the bottom of 96-well plates or 24-well plates. Affymetrix has already made some array types available for this new platform, namely the gene expression products that interrogate that transcriptomes of human, mouse and rat using the IVT kit. DNA-based products such as genotyping and copy number analysis have already been announced to follow

---

[1]Registered trademark of GE Healthcare, Limited, Little Chalfont, United Kingdom.
[2]Registered trademark of Affymetrix, Inc., Santa Clara, California.

in mid-2009. Besides increasing the throughput, the technology also promises to reduce the hands-on time in the lab to 30 minutes as the instrument will do the hybridization, washing and scanning automatically. Due to a smaller chip size, the amount of starting material that is needed has been reduced and the hybridization has been shortened as well. Of course, since this technology has only been launched recently, the impact of the differences in array format, instrument and processing on the data as well as potential adjustments to the data preprocessing has not been investigated outside of Affymetrix.

9.3   Next-gen sequencing: the end for microarrays?

Jay Shendure has published a news and views article with the title "The beginning of the end for microarrays?"[256]. Indeed, with the recent advancements in next-generation sequencing it is becoming very likely that technologies such as the SOLiD platform®[3] (Applied Biosystems), the Genome-Analyzer (Illumina) or the Genome Sequencer FLX®[4] (454 Life Science / Roche)[257] will bring gene expression profiling experiments to a new level as this technology will most likely provide the researcher with more accurate quantitative data and more data per sample overall. Major advantages include:

1. No need to know the sequence of the transcript. With microarrays, it is a prerequisite to know the sequence to be able to design probes. While it is not necessary to have this knowledge when applying sequencing-based strategies, it certainly helps the interpretation of the discovered sequence fragments, especially for novel splice variants and novel transcripts that are not protein-encoding.

2. Transcripts of gene families with high sequence homology are more easily quantifiable as there are no artifacts due to cross-hybridization.

3. Better quantitation of low abundance transcripts. As signal intensities of low abundant transcripts tend to be very close to the level of background, the quantification of such mRNA molecules is very error prone.

4. Theoretically unlimited "dynamic range." Microarray technology typically stores the captured array images in 16 bit TIFF files leading to a theoretically maximum dynamic range of 5 orders of magnitude (16 bit $= 2^{16} = 65{,}536$ levels of gray). For next-generation sequencing technologies, the measurement of the abundance of transcripts is purely

---

[3]Registered trademark of Applied Biosystems, Inc., Foster City, California.
[4]Registered trademark of 454 Life Sciences, Branford, Connecticut.

dependent on the number of sequencing reactions that are being carried out.

While the future looks bright for next-generation sequencing as the tool of choice for gene activity measurements, the purpose of this book is to help researchers in conducting transcriptome analyses today. As with all new technologies, several issues still need to be sorted out before next-generation sequencing can be used on a routine basis. Some of these issues are listed below:

- For some next-gen sequencing technologies read lengths are short as compared to conventional sequencers, preventing *de novo* genome sequence assembly[258]. Three platforms currently have read lengths of approximately 30 bases. Paired-end reads[5] can be used to increase the assembly success.

- The sample preparation procedure can be difficult and can cause low read coverage in A/T rich regions.

- Depending on the application, preparation of libraries can be quite complicated.

- High error rates can occur for some next-gen sequencing technologies, e.g., error that is introduced during sample preparation/sample amplification and dependency of the accuracy for a given base quality value on that base's position in the sequence read[259]. Error rates are especially critical when dealing with short sequence reads.

- The reliabilities of DNA sequence reads cannot always be effectively characterized. Tile images, which constitute the raw data from which DNA sequence information is ultimately derived, can contain erroneous data, e.g., caused by a bubble in the reagents (see Figure 9.1)[260].

- New data types and huge data volumes (10s of terabytes per year for a routinely used instrument need to be archived) pose formidable informatics challenges (e.g., sequence assembly of millions of short-read fragments)[261]. Companies like Illumina try to reduce the problem by

---

[5]Paired-end reads is an approach that is used to order and orient assembled sequences. It essentially means that two DNA sequence reads are separated by an unknown piece of DNA with a known size. Even though the sequence in between is unknown, the size information can be used in the sequence assembly: the two sequenced pieces of DNA (the paired-end reads) may only end up in a region that is separated by this known distance. This way of increasing specificity – a number of short DNA sequences can occur multiple times especially in larger genomes – is often crucial in *de novo* sequencing approaches. However, the power of this approach is tightly linked to the laboratory protocol: the smaller the variability of the size of the inserted unknown sequence between the paired reads, the better the utilization of the paired-end reads for *de novo* assembly.

providing the user with quality scores. This way users are given the option to save just the sequence data together with the quality scores instead of having to store raw images. Still, a strong informatics infrastructure is required to deal with the data volume.

- A consensus is developing that it will soon be cheaper to re-run a sequencing experiment than to store raw images for reanalysis. While this is doable for DNA analyzes, it is less appropriate for gene expression experiments, where a given sample cannot easily be replaced. Experience with Affymetrix technology has proven that reanalysis of old .CEL files with modern analysis technologies can be very beneficial as the accuracy of the conclusions could be increased.

- The explosion in amounts of generated sequence data is challenging the ways data is stored in databases and can time-efficiently be queried with a newly obtained query sequence to test for existing data.

- While technology providers have come up with solutions for building sequence reads, there is still a need for good quality software for sequence mapping.

- There is no golden standard technology. All next generation sequencers use different chemistry than classical sequencing. They furthermore have different strengths and weaknesses, such as longer reads, greater accuracy, or greater data density. In other words, it is likely that instead of one standard approach each technology will have a set of applications where it performs better compared to others.

- Next next-generation sequencing technologies (also called "third-generation sequencers" or "nanopore sequencing") using single-molecule sequencing strategies are already waiting to replace next-generation sequencing (e.g., Pacific Biosciences, Complete Genomics, VisiGen Biotechnologies, US National Human Genome Research Institute in Bethesda, Maryland).

The advantage of the Affymetrix platform today is that many scientists have applied the technology successfully in their research. The research community working on new data analysis techniques and adequate approaches for assessing data quality to focus on the relevant information has caught up and the toolbox is very complete to perform gene expression studies.

The issues faced by next-generation sequencing at this moment are surely solvable, but make this technology difficult to use for routine research. Perhaps after some time, a couple of years, people might eventually start to utilize this technology more routinely when the technology indeed provides better data and saves time and cost, and when software solutions are available to analyze and interpret the data.

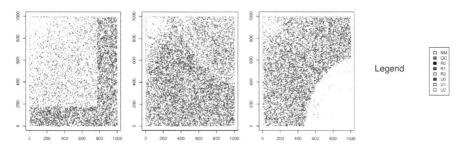

FIGURE 9.1: Errors that can occur on next-gen sequencing tiles. Image reproduced from the article of Dolan and Denver in BMC Bioinformatics[260].

To be absolutely clear about this, the authors have themselves been applying next-generation sequencing successfully in one of the first major scientific papers and are therefore not arguing against this platform. In this publication we identified the mechanism of action of a promising new compound for the treatment of multiple-drug-resistant tuberculosis via 454 Life Sciences sequencing technology. We sequenced three whole bacterial genomes and searched for genes mutated in all three genomes. Comparing the mutations across all genomes identified the one gene that was affected in all strains: the ATP synthase which has now become a promising new target for the development of new compounds against *Mycobacterium tuberculosis*[262].

However, problems like the terabytes of data that are being generated by next-generation sequencers not only pose problems with respect to data archiving (an issue that is critical for pharmaceutical companies for regulatory reasons and to protect intellectual property), but also practical issues such as network speed. As was noted in an editorial in *Nature Methods*, some labs currently carry actual hard drives between buildings to speed up "data transfers"[263]. In general, next-generation sequencing technologies, but also newer generations of Affymetrix microarrays and other technologies that obtain increasingly more measurements per sample are difficult to establish as routine tools without strong support from IT departments.

A likely scenario for the next couple of years is that next-generation sequencing will be used increasingly as a discovery platform for the types of sequences that are available in the RNA, DNA and miRNA world. On the other hand, arrays are likely to be used as a platform to gather quantitative data about the abundance of a given sequence in a sample. This is largely due to the robustness of the platform as well as the familiarity of a large number of scientists and data analysts with the generation and interpretation of microarray data. However, once the costs of creating and using high density arrays are larger than the costs of next-generation sequencing, sequencing will become the method of choice for studying sequence abundance (e.g., gene expression profiling), resequencing, or genomic structural variations. Until then, microarrays will remain the tool of choice for whole genome studies. First studies comparing the data from next-gen sequencing with data from different microarray platforms have shown that there is a high overlap in differentially expressed transcripts especially with Affymetrix microarrays[264].

Irrespective of how the importance of microarrays will develop further, a constantly improving and changing toolbox is part of science. Many of the concepts discussed in this book will remain irrespective of the technology that is applied. Issues like the experimental design, the quality of the samples, the biological question, or the data analysis strategy will remain when a new tool retires current microarray technology. Until then we hope that this book has given you sufficient information to ensure that your gene expression studies using (Affymetrix) microarrays will be conclusive.

Bibliography

[1] Gentleman R, Carey V, Huber W, Irizarry R, Dudoit S, editors: *Bioinformatics and computational biology solutions using R and Bioconductor*. Springer Science, New York, 2005.

[2] Draghici S: *Data analysis tools for DNA microarrays*. CRC Press, 2003.

[3] Team RDC: *R: a language and environment for statistical computing.*. R Foundation for Statistical Computing, Vienna, Austria, 2008.

[4] Naef F, Magnasco MO: **Solving the riddle of the bright mismatches: labeling and effective binding in oligonucleotide arrays.** *Phys Rev E Stat Nonlin Soft Matter Phys* 2003, 68:011906.

[5] Vingron M: **Bioinformatics needs to adopt statistical thinking.** *Bioinformatics* 2001, 17:389–390.

[6] Kapranov P, Cheng J, Dike S, Nix DA, Duttagupta R, Willingham AT, Stadler PF, Hertel J, Hackermüller J, Hofacker IL, Bell I, Cheung E, Drenkow J, Dumais E, Patel S, Helt G, Ganesh M, Ghosh S, Piccolboni A, Sementchenko V, Tammana H, Gingeras TR: **RNA maps reveal new RNA classes and a possible function for pervasive transcription.** *Science* 2007, 316:1484–1488.

[7] Hahne F, Gentleman R: *Bioconductor case studies*. Use R! Springer Science + Business Media, LLC, 2008.

[8] Li X: **ALL: A data package**. Internet, 2006. URL http://bioconductor.org/packages/release/data/experiment/html/ALL.html.

[9] Irizarry RA, Hobbs B, Collin F, Beazer-Barclay YD, Antonellis KJ, Scherf U, Speed TP: **Exploration, normalization, and summaries of high density oligonucleotide array probe level data.** *Biostatistics* 2003, 4:249–264.

[10] Dillman JF, Hege AI, Phillips CS, Orzolek LD, Sylvester AJ, Bossone C, Henemyre-Harris C, Kiser RC, Choi YW, Schlager JJ, Sabourin CL: **Microarray analysis of mouse ear tissue exposed to bis-(2-chloroethyl) sulfide: gene expression profiles correlate with treatment efficacy and an established clinical endpoint.** *J Pharmacol Exp Ther* 2006, 317:76–87.

[11] Gouze JN, Gouze E, Popp MP, Bush ML, Dacanay EA, Kay JD, Levings PP, Patel KR, Saran JPS, Watson RS, Ghivizzani SC: **Exogenous glucosamine globally protects chondrocytes from the arthritogenic effects of IL-1beta.** *Arthritis Res Ther* 2006, 8:R173.

[12] Hodges A, Hughes G, Brooks S, Elliston L, Holmans P, Dunnett SB, Jones L: **Brain gene expression correlates with changes in behavior in the R6/1 mouse model of Huntington's disease.** *Genes Brain Behav* 2007, 7:288–299.

[13] Sørensen JG, Nielsen MM, Kruhøffer M, Justesen J, Loeschcke V: **Full genome gene expression analysis of the heat stress response in Drosophila melanogaster.** *Cell Stress Chaperones* 2005, 10:312–328.

[14] Als AB, Dyrskjøt L, von der Maase H, Koed K, Mansilla F, Toldbod HE, Jensen JL, Ulhøi BP, Sengeløv L, Jensen KME, Orntoft TF: **Emmprin and survivin predict response and survival following cisplatin-containing chemotherapy in patients with advanced bladder cancer.** *Clin Cancer Res* 2007, 13:4407–4414.

[15] Canales RD, Luo Y, Willey JC, Austermiller B, Barbacioru CC, Boysen C, Hunkapiller K, Jensen RV, Knight CR, Lee KY, Ma Y, Maqsodi B, Papallo A, Peters EH, Poulter K, Ruppel PL, Samaha RR, Shi L, Yang W, Zhang L, Goodsaid FM: **Evaluation of DNA microarray results with quantitative gene expression platforms.** *Nat Biotechnol* 2006, 24:1115–1122.

[16] Perez-Iratxeta C, Andrade MA: **Inconsistencies over time in 5% of NetAffx probe-to-gene annotations.** *BMC Bioinformatics* 2005, 6:183.

[17] Lopez AJ: **Alternative splicing of pre-mRNA: developmental consequences and mechanisms of regulation.** *Annu Rev Genet* 1998, 32:279–305.

[18] Affymetrix: **NetAffx**. Internet. URL https://www.affymetrix.com/analysis/netaffx/index.affx.

[19] Dai M, Wang P, Boyd AD, Kostov G, Athey B, Jones EG, Bunney WE, Myers RM, Speed TP, Akil H, Watson SJ, Meng F: **Evolving gene/transcript definitions significantly alter the interpretation of GeneChip data.** *Nucleic Acids Res* 2005, 33:e175.

[20] Microarray Lab UoM: **Custom CDF**. Internet. URL http://brainarray.mbni.med.umich.edu/Brainarray/Database/CustomCDF/genomic_curated_CDF.asp.

[21] Sandberg R, Larsson O: **Improved precision and accuracy for microarrays using updated probe set definitions.** *BMC Bioinformatics* 2007, 8:48.

[22] Wheeler DL, Church DM, Edgar R, Federhen S, Helmberg W, Madden TL, Pontius JU, Schuler GD, Schriml LM, Sequeira E, Suzek TO, Tatusova TA, Wagner L: **Database resources of the National Center for Biotechnology Information: update.** *Nucleic Acids Res* 2004, 32:D35–D40.

[23] Benson DA, Karsch-Mizrachi I, Lipman DJ, Ostell J, Wheeler DL: **GenBank.** *Nucleic Acids Res* 2005, 33:D34–D38.

[24] Boguski MS, Lowe TM, Tolstoshev CM: **dbEST–database for "expressed sequence tags".** *Nat Genet* 1993, 4:332–333.

[25] Pruitt KD, Tatusova T, Maglott DR: **NCBI Reference Sequence (RefSeq): a curated non-redundant sequence database of genomes, transcripts and proteins.** *Nucleic Acids Res* 2005, 33:D501–D504.

[26] Johnson JM, Castle J, Garrett-Engele P, Kan Z, Loerch PM, Armour CD, Santos R, Schadt EE, Stoughton R, Shoemaker DD: **Genome-wide survey of human alternative pre-mRNA splicing with exon junction microarrays.** *Science* 2003, 302:2141–2144.

[27] Ledford H: **Human genes are multitaskers.** *Nature* 2008, 456:9.

[28] Burge C, Karlin S: **Prediction of complete gene structures in human genomic DNA.** *J Mol Biol* 1997, 268:78–94.

[29] Korf I, Flicek P, Duan D, Brent MR: **Integrating genomic homology into gene structure prediction.** *Bioinformatics* 2001, 17 Suppl 1:S140–S148.

[30] Parra G, Blanco E, Guigó R: **GeneID in Drosophila.** *Genome Res* 2000, 10:511–515.

[31] Robinson MD, Speed TP: **A comparison of Affymetrix gene expression arrays.** *BMC Bioinformatics* 2007, 8:449.

[32] Kapranov P, Willingham AT, Gingeras TR: **Genome-wide transcription and the implications for genomic organization.** *Nat Rev Genet* 2007, 8:413–423.

[33] Dumur CI, Nasim S, Best AM, Archer KJ, Ladd AC, Mas VR, Wilkinson DS, Garrett CT, Ferreira-Gonzalez A: **Evaluation of quality-control criteria for microarray gene expression analysis.** *Clin Chem* 2004, 50:1994–2002.

[34] Shi L, Reid LH, Jones WD, Shippy R, Warrington JA, Baker SC, Collins PJ, de Longueville F, Kawasaki ES, Lee KY, Luo Y, Sun YA, Willey JC, Setterquist RA, Fischer GM, Tong W, Dragan YP, Dix DJ, Frueh FW, Goodsaid FM, Herman D, Jensen RV, Johnson CD, Lobenhofer EK, Puri RK, Scherf U, Thierry-Mieg J, Wang C, Wilson M, Wolber

PK, Zhang L, Amur S, Bao W, Barbacioru CC, Lucas AB, Bertholet V, Boysen C, Bromley B, Brown D, Brunner A, Canales R, Cao XM, Cebula TA, Chen JJ, Cheng J, Chu TM, Chudin E, Corson J, Corton JC, Croner LJ, Davies C, Davison TS, Delenstarr G, Deng X, Dorris D, Eklund AC, Fan XH, Fang H, Fulmer-Smentek S, Fuscoe JC, Gallagher K, Ge W, Guo L, Guo X, Hager J, Haje PK, Han J, Han T, Harbottle HC, Harris SC, Hatchwell E, Hauser CA, Hester S, Hong H, Hurban P, Jackson SA, Ji H, Knight CR, Kuo WP, Leclerc JE, Levy S, Li QZ, Liu C, Liu Y, Lombardi MJ, Ma Y, Magnuson SR, Maqsodi B, McDaniel T, Mei N, Myklebost O, Ning B, Novoradovskaya N, Orr MS, Osborn TW, Papallo A, Patterson TA, Perkins RG, Peters EH, Peterson R, Philips KL, Pine PS, Pusztai L, Qian F, Ren H, Rosen M, Rosenzweig BA, Samaha RR, Schena M, Schroth GP, Shchegrova S, Smith DD, Staedtler F, Su Z, Sun H, Szallasi Z, Tezak Z, Thierry-Mieg D, Thompson KL, Tikhonova I, Turpaz Y, Vallanat B, Van C, Walker SJ, Wang SJ, Wang Y, Wolfinger R, Wong A, Wu J, Xiao C, Xie Q, Xu J, Yang W, Zhang L, Zhong S, Zong Y, Slikker W: **The MicroArray Quality Control (MAQC) project shows inter- and intraplatform reproducibility of gene expression measurements.** *Nat Biotechnol* 2006, 24:1151–1161.

[35] Editorial: **Making the most of microarrays.** *Nat Biotechnol* 2006, 24:1039.

[36] Tian Q, Stepaniants SB, Mao M, Weng L, Feetham MC, Doyle MJ, Yi EC, Dai H, Thorsson V, Eng J, Goodlett D, Berger JP, Gunter B, Linseley PS, Stoughton RB, Aebersold R, Collins SJ, Hanlon WA, Hood LE: **Integrated genomic and proteomic analyses of gene expression in Mammalian cells.** *Mol Cell Proteomics* 2004, 3:960–969.

[37] Li S, Pozhitkov A, Brouwer M: **A competitive hybridization model predicts probe signal intensity on high density DNA microarrays.** *Nucleic Acids Res* 2008, 36:6585–6591.

[38] Held GA, Grinstein G, Tu Y: **Relationship between gene expression and observed intensities in DNA microarrays–a modeling study.** *Nucleic Acids Res* 2006, 34:e70.

[39] Peeters PJ, Fierens FLP, van den Wyngaert I, Goehlmann HW, Swagemakers SM, Kass SU, Langlois X, Pullan S, Stenzel-Poore MP, Steckler T: **Gene expression profiles highlight adaptive brain mechanisms in corticotropin releasing factor overexpressing mice.** *Brain Res Mol Brain Res* 2004, 129:135–150.

[40] Gallitano-Mendel A, Izumi Y, Tokuda K, Zorumski CF, Howell MP, Muglia LJ, Wozniak DF, Milbrandt J: **The immediate early gene**

early growth response gene 3 mediates adaptation to stress and novelty. *Neuroscience* 2007, 148:633–643.

[41] Rainen L, Oelmueller U, Jurgensen S, Wyrich R, Ballas C, Schram J, Herdman C, Bankaitis-Davis D, Nicholls N, Trollinger D, Tryon V: **Stabilization of mRNA expression in whole blood samples.** *Clin Chem* 2002, 48:1883–1890.

[42] Field LA, Jordan RM, Hadix JA, Dunn MA, Shriver CD, Ellsworth RE, Ellsworth DL: **Functional identity of genes detectable in expression profiling assays following globin mRNA reduction of peripheral blood samples.** *Clin Biochem* 2007, 40:499–502.

[43] Debey S, Schoenbeck U, Hellmich M, Gathof BS, Pillai R, Zander T, Schultze JL: **Comparison of different isolation techniques prior to gene expression profiling of blood derived cells: impact on physiological responses, on overall expression and the role of different cell types.** *Pharmacogenomics J* 2004, 4:193–207.

[44] Tomita H, Vawter MP, Walsh DM, Evans SJ, Choudary PV, Li J, Overman KM, Atz ME, Myers RM, Jones EG, Watson SJ, Akil H, Bunney WE: **Effect of agonal and postmortem factors on gene expression profile: quality control in microarray analyses of postmortem human brain.** *Biol Psychiatry* 2004, 55:346–352.

[45] Popova T, Mennerich D, Weith A, Quast K: **Effect of RNA quality on transcript intensity levels in microarray analysis of human post-mortem brain tissues.** *BMC Genomics* 2008, 9:91.

[46] Blow N: **Tissue preparation: Tissue issues.** *Nature* 2007, 448:959–963.

[47] Penland SK, Keku TO, Torrice C, He X, Krishnamurthy J, Hoadley KA, Woosley JT, Thomas NE, Perou CM, Sandler RS, Sharpless NE: **RNA expression analysis of formalin-fixed paraffin-embedded tumors.** *Lab Invest* 2007, 87:383–391.

[48] Hughes P, Marshall D, Reid Y, Parkes H, Gelber C: **The costs of using unauthenticated, over-passaged cell lines: how much more data do we need?** *Biotechniques* 2007, 43:575, 577–8, 581–2 passim.

[49] Weinstein IB: **Cancer. Addiction to oncogenes–the Achilles heal of cancer.** *Science* 2002, 297:63–64.

[50] Lein ES, Hawrylycz MJ, Ao N, Ayres M, Bensinger A, Bernard A, Boe AF, Boguski MS, Brockway KS, Byrnes EJ, Chen L, Chen L, Chen TM, Chin MC, Chong J, Crook BE, Czaplinska A, Dang CN, Datta S, Dee NR, Desaki AL, Desta T, Diep E, Dolbeare TA, Donelan MJ, Dong HW, Dougherty JG, Duncan BJ, Ebbert AJ, Eichele G, Estin LK, Faber C, Facer BA, Fields R, Fischer SR, Fliss TP, Frensley C, Gates

SN, Glattfelder KJ, Halverson KR, Hart MR, Hohmann JG, Howell MP, Jeung DP, Johnson RA, Karr PT, Kawal R, Kidney JM, Knapik RH, Kuan CL, Lake JH, Laramee AR, Larsen KD, Lau C, Lemon TA, Liang AJ, Liu Y, Luong LT, Michaels J, Morgan JJ, Morgan RJ, Mortrud MT, Mosqueda NF, Ng LL, Ng R, Orta GJ, Overly CC, Pak TH, Parry SE, Pathak SD, Pearson OC, Puchalski RB, Riley ZL, Rockett HR, Rowland SA, Royall JJ, Ruiz MJ, Sarno NR, Schaffnit K, Shapovalova NV, Sivisay T, Slaughterbeck CR, Smith SC, Smith KA, Smith BI, Sodt AJ, Stewart NN, Stumpf KR, Sunkin SM, Sutram M, Tam A, Teemer CD, Thaller C, Thompson CL, Varnam LR, Visel A, Whitlock RM, Wohnoutka PE, Wolkey CK, Wong VY, Wood M, Yaylaoglu MB, Young RC, Youngstrom BL, Yuan XF, Zhang B, Zwingman TA, Jones AR: **Genome-wide atlas of gene expression in the adult mouse brain.** *Nature* 2007, 445:168–176.

[51] Allen Institute for Brain Science: **Allen Brain Atlas (c) 2004-2007.** Internet. URL http://www.brain-map.org.

[52] Champy M, Selloum M, Piard L, Zeitler V, Caradec C, Chambon P, Auwerx J: **Mouse functional genomics requires standardization of mouse handling and housing conditions.** *Mammalian Genome* 2004, 15:768–783.

[53] Kozul CD, Nomikos AP, Hampton TH, Warnke LA, Gosse JA, Davey JC, Thorpe JE, Jackson BP, Ihnat MA, Hamilton JW: **Laboratory diet profoundly alters gene expression and confounds genomic analysis in mouse liver and lung.** *Chem Biol Interact* 2008, 173:129–140.

[54] Weiss B, Stern S, Cernichiari E, Gelein R: **Methylmercury contamination of laboratory animal diets.** *Environ Health Perspect* 2005, 113:1120–1122.

[55] Thompson KL, Pine PS, Rosenzweig BA, Turpaz Y, Retief J: **Characterization of the effect of sample quality on high density oligonucleotide microarray data using progressively degraded rat liver RNA.** *BMC Biotechnol* 2007, 7:57.

[56] Gelder RNV, von Zastrow ME, Yool A, Dement WC, Barchas JD, Eberwine JH: **Amplified RNA synthesized from limited quantities of heterogeneous cDNA.** *Proc Natl Acad Sci U S A* 1990, 87:1663–1667.

[57] Cope L, Hartman SM, Göhlmann HWH, Tiesman JP, Irizarry RA: **Analysis of Affymetrix GeneChip data using amplified RNA.** *Biotechniques* 2006, 40:165–166, 168, 170.

[58] Klebanov L, Yakovlev A: **How high is the level of technical noise in microarray data?** *Biol Direct* 2007, 2:9.

[59] Pritchard CC, Hsu L, Delrow J, Nelson PS: **Project normal: defining normal variance in mouse gene expression.** *Proc Natl Acad Sci U S A* 2001, 98:13266–13271.

[60] Kendziorski C, Irizarry RA, Chen KS, Haag JD, Gould MN: **On the utility of pooling biological samples in microarray experiments.** *Proc Natl Acad Sci U S A* 2005, 102:4252–4257.

[61] Shih JH, Michalowska AM, Dobbin K, Ye Y, Qiu TH, Green JE: **Effects of pooling mRNA in microarray class comparisons.** *Bioinformatics* 2004, 20:3318–3325.

[62] Mary-Huard T, Daudin JJ, Baccini M, Biggeri A, Bar-Hen A: **Biases induced by pooling samples in microarray experiments.** *Bioinformatics* 2007, 23:i313–i318.

[63] Heid CA, Stevens J, Livak KJ, Williams PM: **Real time quantitative PCR.** *Genome Res* 1996, 6:986–994.

[64] Willey JC, Crawford EL, Knight CR, Warner KA, Motten CA, Herness EA, Zahorchak RJ, Graves TG: **Standardized RT-PCR and the standardized expression measurement center.** *Methods Mol Biol* 2004, 258:13–41.

[65] Flagella M, Bui S, Zheng Z, Nguyen CT, Zhang A, Pastor L, Ma Y, Yang W, Crawford KL, McMaster GK, Witney F, Luo Y: **A multiplex branched DNA assay for parallel quantitative gene expression profiling.** *Anal Biochem* 2006, 352:50–60.

[66] Tong W, Lucas AB, Shippy R, Fan X, Fang H, Hong H, Orr MS, Chu TM, Guo X, Collins PJ, Sun YA, Wang SJ, Bao W, Wolfinger RD, Shchegrova S, Guo L, Warrington JA, Shi L: **Evaluation of external RNA controls for the assessment of microarray performance.** *Nat Biotechnol* 2006, 24:1132–1139.

[67] Jones L, Goldstein DR, Hughes G, Strand AD, Collin F, Dunnett SB, Kooperberg C, Aragaki A, Olson JM, Augood SJ, Faull RLM, Luthi-Carter R, Moskvina V, Hodges AK: **Assessment of the relationship between pre-chip and post-chip quality measures for Affymetrix GeneChip expression data.** *BMC Bioinformatics* 2006, 7:211.

[68] Holloway A, Oshlack A, Diyagama D, Bowtell D, Smyth G: **Statistical analysis of an RNA titration series evaluates microarray precision and sensitivity on a whole-array basis.** *BMC Bioinformatics* 2006, 7:511.

[69] Shippy R, Fulmer-Smentek S, Jensen RV, Jones WD, Wolber PK, Johnson CD, Pine PS, Boysen C, Guo X, Chudin E, Sun YA, Willey JC, Thierry-Mieg J, Thierry-Mieg D, Setterquist RA, Wilson M, Lucas AB,

Novoradovskaya N, Papallo A, Turpaz Y, Baker SC, Warrington JA, Shi L, Herman D: **Using RNA sample titrations to assess microarray platform performance and normalization techniques.** *Nat Biotechnol* 2006, 24:1123–1131.

[70] Huber W, von Heydebreck A, Sültmann H, Poustka A, Vingron M: **Variance stabilization applied to microarray data calibration and to the quantification of differential expression.** *Bioinformatics* 2002, 18 Suppl 1:S96–104.

[71] Wu Z, Irizarry RA, Gentleman R, Murillo FM, Spencer F: **A model based background adjustment for oligonucleotide expression arrays.** *Johns Hopkins University, Dept of Biostatistics Working Papers* 2004, Working Paper 1:1–26.

[72] Katz S, Irizarry RA, Lin X, Tripputi M, Porter MW: **A summarization approach for Affymetrix GeneChip data using a reference training set from a large, biologically diverse database.** *BMC Bioinformatics* 2006, 7:464.

[73] Li C, Wong WH: **Model-based analysis of oligonucleotide arrays: expression index computation and outlier detection.** *Proc Natl Acad Sci U S A* 2001, 98:31–36.

[74] Harrison AP, Johnston CE, Orengo CA: **Establishing a major cause of discrepancy in the calibration of Affymetrix GeneChips.** *BMC Bioinformatics* 2007, 8:195.

[75] Hochreiter S, Clevert DA, Obermayer K: **A new summarization method for Affymetrix probe level data.** *Bioinformatics* 2006, 22:943–949.

[76] Choe SE, Boutros M, Michelson AM, Church GM, Halfon MS: **Preferred analysis methods for Affymetrix GeneChips revealed by a wholly defined control dataset.** *Genome Biol* 2005, 6:R16.

[77] Gentleman P, Ding B, Dudoit S: *Bioinformatics and computational biology solutions using R and Bioconductor.* Springer, 2005, 192–194.

[78] Bolstad B: *Low Level Analysis of High-density Oligonucleotide Array Data: Background, Normalization and Summarization.* Ph.D. thesis, University of California, Berkeley, 2004.

[79] Rodgers JL, Nicewander WA: **Thirteen ways to look at the correlation coefficient.** *The American Statistician* 1988, 42:59–66.

[80] Anscombe FJ: **Graphs in statistical analysis.** *The American Statistician* 1973, 27:17–21.

[81] Magari RT: **Statistics for laboratory method comparison studies.** *BioPharm* 2002, 15:28–32.

[82] I-Kuei Lin L: **A concordance correlation coefficient to evaluate reproducibility.** *Biometrics* 1989, 45:255–268.

[83] Nickerson CAE: **A note on "a concordance correlation coefficient to evaluate reproducibility".** *Biometrics* 1997, 53:1503–1507.

[84] Müller R, Büttner P: **A critical discussion of intraclass correlation coefficients.** *Stat Med* 1994, 13:2465–2476.

[85] Aerssens J, Camilleri M, Talloen W, Thielemans L, Göhlmann HWH, Wyngaert IVD, Thielemans T, Hoogt RD, Andrews CN, Bharucha AE, Carlson PJ, Busciglio I, Burton DD, Smyrk T, Urrutia R, Coulie B: **Alterations in mucosal immunity identified in the colon of patients with irritable bowel syndrome.** *Clin Gastroenterol Hepatol* 2008, 6:194–205.

[86] Johnson WE, Li C, Rabinovic A: **Adjusting batch effects in microarray expression data using empirical Bayes methods.** *Biostatistics* 2007, 8:118–127.

[87] Slonim DK: **From patterns to pathways: gene expression data analysis comes of age.** *Nat Genet* 2002, 32 Suppl:502–508.

[88] Pearson H: **Double check casts doubt on statistics in published papers.** *Nature* 2004, 429:490.

[89] Cates: **Microarray experiments.** Internet - Connexions, 2005. URL http://cnx.org/content/m11050/2.16/.

[90] Vandenbroeck P, Wouters L, Molenberghs G, Gestel JV, Bijnens L: **Teaching statistical thinking to life scientists: a case-based approach.** *J Biopharm Stat* 2006, 16:61–75.

[91] Breiman L: **Statistical modeling: The two cultures.** *Stat Sci* 2001, 16:199–231.

[92] Rao CR: **R.A. Fisher: The founder of modern statistics.** *Statistical Science* 1992, 7:34–48.

[93] Mansmann U, Ruschhaupt M, Huber W: **Reproducible statistical analysis in microarray profiling studies.** *Methods Inf Med* 2006, 45:139–145.

[94] Tibshirani RJ, Efron B: **Pre-validation and inference in microarrays.** *Stat Appl Genet Mol Biol* 2002, 1:Article1.

[95] van 't Veer LJ, Dai H, van de Vijver MJ, He YD, Hart AAM, Mao M, Peterse HL, van der Kooy K, Marton MJ, Witteveen AT, Schreiber GJ, Kerkhoven RM, Roberts C, Linsley PS, Bernards R, Friend SH: **Gene expression profiling predicts clinical outcome of breast cancer.** *Nature* 2002, 415:530–536.

[96] Calza S, Raffelsberger W, Ploner A, Sahel J, Leveillard T, Pawitan Y: **Filtering genes to improve sensitivity in oligonucleotide microarray data analysis.** *Nucleic Acids Res* 2007, 35:e102.

[97] Talloen W, Clevert DA, Hochreiter S, Amaratunga D, Bijnens L, Kass S, Göhlmann HWH: **I/NI-calls for the exclusion of non-informative genes: a highly effective filtering tool for microarray data.** *Bioinformatics* 2007, 23:2897–2902.

[98] Su AI, Cooke MP, Ching KA, Hakak Y, Walker JR, Wiltshire T, Orth AP, Vega RG, Sapinoso LM, Moqrich A, Patapoutian A, Hampton GM, Schultz PG, Hogenesch JB: **Large-scale analysis of the human and mouse transcriptomes.** *Proc Natl Acad Sci U S A* 2002, 99:4465–4470.

[99] Jongeneel CV, Iseli C, Stevenson BJ, Riggins GJ, Lal A, Mackay A, Harris RA, O'Hare MJ, Neville AM, Simpson AJG, Strausberg RL: **Comprehensive sampling of gene expression in human cell lines with massively parallel signature sequencing.** *Proc Natl Acad Sci U S A* 2003, 100:4702–4705.

[100] m Liu W, Mei R, Di X, Ryder TB, Hubbell E, Dee S, Webster TA, Harrington CA, h Ho M, Baid J, Smeekens SP: **Analysis of high density expression microarrays with signed-rank call algorithms.** *Bioinformatics* 2002, 18:1593–1599.

[101] McClintick JN, Edenberg HJ: **Effects of filtering by Present call on analysis of microarray experiments.** *BMC Bioinformatics* 2006, 7:49.

[102] Pepper SD, Saunders EK, Edwards LE, Wilson CL, Miller CJ: **The utility of MAS5 expression summary and detection call algorithms.** *BMC Bioinformatics* 2007, 8:273.

[103] Butte A: **The use and analysis of microarray data.** *Nat Rev Drug Discov* 2002, 1:951–960.

[104] Clarke R, Ressom HW, Wang A, Xuan J, Liu MC, Gehan EA, Wang Y: **The properties of high-dimensional data spaces: implications for exploring gene and protein expression data.** *Nat Rev Cancer* 2008, 8:37–49.

[105] Garge N, Page G, Sprague A, Gorman B, Allison D: **Reproducible clusters from microarray research: whither?** *BMC bioinformatics* 2005, 6:S10.

[106] Bair E, Tibshirani R: **Semi-supervised methods to predict patient survival from gene expression data.** *PLoS Biol* 2004, 2:E108.

[107] Hastie T, Tibshirani R, Eisen MB, Alizadeh A, Levy R, Staudt L, Chan WC, Botstein D, Brown P: **"Gene shaving" as a method for iden-**

tifying distinct sets of genes with similar expression patterns. *Genome Biol* 2000, 1:RESEARCH0003.

[108] Ward JH: **Hierarchical grouping to optimize an objective function.** *Journal of the American Statistical Association* 1963, 58:236–244.

[109] D'haeseleer P: **How does gene expression clustering work?** *Nat Biotechnol* 2005, 23:1499–1501.

[110] Thiel E, Kranz BR, Raghavachar A, Bartram CR, Löffler H, Messerer D, Ganser A, Ludwig WD, Büchner T, Hoelzer D: **Prethymic phenotype and genotype of pre-T (CD7+/ER-)-cell leukemia and its clinical significance within adult acute lymphoblastic leukemia.** *Blood* 1989, 73:1247–1258.

[111] Chiaretti S, Li X, Gentleman R, Vitale A, Vignetti M, Mandelli F, Ritz J, Foa R: **Gene expression profile of adult T-cell acute lymphocytic leukemia identifies distinct subsets of patients with different response to therapy and survival.** *Blood* 2004, 103:2771–2778.

[112] Goeman JJ, Bühlmann P: **Analyzing gene expression data in terms of gene sets: methodological issues.** *Bioinformatics* 2007, 23:980–987.

[113] Eisen MB, Spellman PT, Brown PO, Botstein D: **Cluster analysis and display of genome-wide expression patterns.** *Proc Natl Acad Sci U S A* 1998, 95:14863–14868.

[114] Kaufman L, Rousseeuw: *Finding groups in data: An introduction to cluster analysis.* Wiley, New York, 1990.

[115] Rahnenführer J: **Clustering algorithms and other exploratory methods for microarray data analysis.** *Methods Inf Med* 2005, 44:444–448.

[116] MacQueen J: *Proceedings of the Fifth Berkeley Symposium on Mathematical Statistics and Probability. Volume I, Statistics..* University of California Press, 1967, .

[117] Kohonen T: **The self-organizing map.** *Proceedings of the IEEE* 1990, 78:1464–1480.

[118] Tamayo P, Slonim D, Mesirov J, Zhu Q, Kitareewan S, Dmitrovsky E, Lander ES, Golub TR: **Interpreting patterns of gene expression with self-organizing maps: methods and application to hematopoietic differentiation.** *Proc Natl Acad Sci U S A* 1999, 96:2907–2912.

[119] Thalamuthu A, Mukhopadhyay I, Zheng X, Tseng GC: **Evaluation and comparison of gene clustering methods in microarray analysis.** *Bioinformatics* 2006, 22:2405–2412.

[120] Dudoit S, Fridlyand J: **A prediction-based resampling method for estimating the number of clusters in a dataset.** *Genome Biol* 2002, 3:RESEARCH0036.

[121] Medvedovic M, Yeung KY, Bumgarner RE: **Bayesian mixture model based clustering of replicated microarray data.** *Bioinformatics* 2004, 20:1222–1232.

[122] Fraley C, Raftery AE: **Model-based clustering, discriminant analysis, and density estimation.** *Journal of the American Statistical Association* 2002, 97:611–631.

[123] Tseng GC, Wong WH: **Tight clustering: a resampling-based approach for identifying stable and tight patterns in data.** *Biometrics* 2005, 61:10–16.

[124] Pan W, Lin J, Le CT: **Model-based cluster analysis of microarray gene-expression data.** *Genome Biol* 2002, 3:RESEARCH0009.

[125] Madeira SC, Oliveira AL: **Biclustering algorithms for biological data analysis: a survey.** *IEEE/ACM Trans Comput Biol Bioinform* 2004, 1:24–45.

[126] Turner HL, Bailey TC, Krzanowski WJ, Hemingway CA: **Biclustering models for structured microarray data.** *IEEE/ACM Trans Comput Biol Bioinform* 2005, 2:316–329.

[127] Cheng Y, Church GM: **Biclustering of expression data.** *Proc Int Conf Intell Syst Mol Biol* 2000, 8:93–103.

[128] Cheng KO, Law NF, Siu WC, Liew AWC: **Identification of coherent patterns in gene expression data using an efficient biclustering algorithm and parallel coordinate visualization.** *BMC Bioinformatics* 2008, 9:210.

[129] Sheng Q, Moreau Y, Moor BD: **Biclustering microarray data by Gibbs sampling.** *Bioinformatics* 2003, 19 Suppl 2:ii196–ii205.

[130] Prelic A, Bleuler S, Zimmermann P, Wille A, Bühlmann P, Gruissem W, Hennig L, Thiele L, Zitzler E: **A systematic comparison and evaluation of biclustering methods for gene expression data.** *Bioinformatics* 2006, 22:1122–1129.

[131] Murali TM, Kasif S: **Extracting conserved gene expression motifs from gene expression data.** *Pacific Symposium on Biocomputing* 2003, 8:77–88.

[132] Ihmels J, Friedlander G, Bergmann S, Sarig O, Ziv Y, Barkai N: **Revealing modular organization in the yeast transcriptional network.** *Nat Genet* 2002, 31:370–377.

[133] Ihmels J, Bergmann S, Barkai N: **Defining transcription modules using large-scale gene expression data.** *Bioinformatics* 2004, 20:1993–2003.

[134] Santamaría R, Therón R, Quintales L: **A visual analytics approach for understanding biclustering results from microarray data.** *BMC Bioinformatics* 2008, 9:247.

[135] Pearson K: **On lines and planes of closest fit to systems of points in space.** *Philosophical Magazine* 1901, 2:559–572.

[136] Fellenberg K, Hauser NC, Brors B, Neutzner A, Hoheisel JD, Vingron M: **Correspondence analysis applied to microarray data.** *Proc Natl Acad Sci U S A* 2001, 98:10781–10786.

[137] Greenacre M: **Correspondence analysis in medical research.** *Stat Methods Med Res* 1992, 1:97–117.

[138] Lewi PJ: **The use of multivariate statistics in industrial pharmacology.** *Pharmacol Ther [B]* 1978, 3:481–537.

[139] Wouters L, Göhlmann HW, Bijnens L, Kass SU, Molenberghs G, Lewi PJ: **Graphical exploration of gene expression data: a comparative study of three multivariate methods.** *Biometrics* 2003, 59:1131–1139.

[140] Johnson RA, Wichern DW: *Applied multivariate statistical analysis. 4th Edition.* Prentice-Hall, 1998.

[141] van de Wetering M, Oosterwegel M, Holstege F, Dooyes D, Suijkerbuijk R, van Kessel AG, Clevers H: **The human T cell transcription factor-1 gene. Structure, localization, and promoter characterization.** *J Biol Chem* 1992, 267:8530–8536.

[142] Krissansen GW, Owen MJ, Verbi W, Crumpton MJ: **Primary structure of the T3 gamma subunit of the T3/T cell antigen receptor complex deduced from cDNA sequences: evolution of the T3 gamma and delta subunits.** *EMBO J* 1986, 5:1799–1808.

[143] Smyth GK, Yang YH, Speed T: **Statistical issues in cDNA microarray data analysis.** *Methods Mol Biol* 2003, 224:111–136.

[144] Witten D, Tibshirani R: **A comparison of fold change and the t-statistic for microarray data analysis.** Technical report, Stanford University, 2007. URL http://www-stat.stanford.edu/~tibs/ftp/FCTComparison.pdf.

[145] Jeffery IB, Higgins DG, Culhane AC: **Comparison and evaluation of methods for generating differentially expressed gene lists from microarray data.** *BMC Bioinformatics* 2006, 7:359.

[146] Kooperberg C, Aragaki A, Strand AD, Olson JM: **Significance testing for small microarray experiments.** *Stat Med* 2005, 24:2281–2298.

[147] Astrand M, Mostad P, Rudemo M: **Empirical Bayes models for multiple probe type microarrays at the probe level.** *BMC Bioinformatics* 2008, 9:156.

[148] Shi L, Tong W, Fang H, Scherf U, Han J, Puri RK, Frueh FW, Goodsaid FM, Guo L, Su Z, Han T, Fuscoe JC, Xu ZA, Patterson TA, Hong H, Xie Q, Perkins RG, Chen JJ, Casciano DA: **Cross-platform comparability of microarray technology: intra-platform consistency and appropriate data analysis procedures are essential.** *BMC Bioinformatics* 2005, 6 Suppl 2:S12.

[149] Jenny D: **BioConductor forum discussion on fold change versus significance testing.** URL http://thread.gmane.org/gmane.science.biology.informatics.conductor/18765/focus=18848.

[150] Tusher VG, Tibshirani R, Chu G: **Significance analysis of microarrays applied to the ionizing radiation response.** *Proc Natl Acad Sci U S A* 2001, 98:5116–5121.

[151] Efron B, Tibshirani R, Storey J, Tusher V: **Empirical Bayes analysis of a microarray experiment.** *Journal of the American Statistical Association* 2001, 96:1151–1160.

[152] Broberg P: **Statistical methods for ranking differentially expressed genes.** *Genome Biol* 2003, 4:R41.

[153] Smyth GK: **Linear models and empirical bayes methods for assessing differential expression in microarray experiments.** *Stat Appl Genet Mol Biol* 2004, 3:Article3.

[154] Lönnstedt I, Speed T: **Replicated microarray data.** *Statistica Sinica* 2002, 12:31–46.

[155] Baldi P, Long AD: **A Bayesian framework for the analysis of microarray expression data: regularized t -test and statistical inferences of gene changes.** *Bioinformatics* 2001, 17:509–519.

[156] Smyth GK: *Bioinformatics and computational biology solutions using R and Bioconductor.* Springer, 2005, 397–420.

[157] Cui X, Churchill GA: **Statistical tests for differential expression in cDNA microarray experiments.** *Genome Biol* 2003, 4:210.

[158] Steinhoff C, Vingron M: **Normalization and quantification of differential expression in gene expression microarrays.** *Brief Bioinform* 2006, 7:166–177.

[159] Trochim: **General linear model**, 2006. URL http://www.socialresearchmethods.net/kb/genlin.php.

[160] Van de Geer S, Van Houwelingen H: **High-dimensional data: p much larger than n in mathematical statistics and bio-medical applications.** *Bernoulli* 2004, 10:939–942.

[161] Hoerl A, Kennard R: **Ridge regression: applications to nonorthogonal problems.** *Technometrics* 1970, 12:69–82.

[162] Verweij P, Van Houwelingen H: **Penalized likelihood in Cox regression.** *Statistics in medicine* 1994, 13:2427–2436.

[163] Boyd S, Vandenberghe L: *Convex optimization.* Cambridge University Press, 2004.

[164] Hastie T, Tibshirani R, Friedman J: *The elements of statistical learning.* Springer Verlag, New York, US, 2001.

[165] Tibshirani R: **Regression shrinkage and selection via the Lasso.** *Journal of the Royal Statistical Society Series B (Methodological)* 1996, 58:267–288.

[166] Efron B, Hastie T, Johnstone I, Tibshirani R: **Least angle regression.** *The Annals of Statistics* 2004, 32:407–451.

[167] Friedman J, Popescu B: **Gradient directed regularization for linear regression and classification.** Technical report, Stanford University, 2004. URL http://www-stat. stanford. edu/jhf/ftp/path. pdf.

[168] Zou H, Hastie T: **Regularization and variable selection via the elastic net.** *Journal of the Royal Statistical Society Series B* 2005, 67:301–320.

[169] Goeman J: **L1 and L2 penalized regression models**. Vignette R Package Penalized. URL http://cran.nedmirror.nl/web/packages/penalized/vignettes/penalized.pdf.

[170] StatSoft I: *Electronic statistics textbook.* StatSoft, 1984.

[171] Satterthwaite F: **An approximate distribution of estimates of variance components.** *Biometrics Bulletin* 1946, 2:110–114.

[172] Holm S: **A simple sequentially rejective multiple test procedure.** *Scandinavian Journal of Statistics* 1979, 6:65–70.

[173] Benjamini Y, Hochberg Y: **Controlling the false discovery rate: a practical and powerful approach to multiple testing.** *Journal Royal Statistical Society Series B* 1995, 57:289–300.

[174] Benjamini Y, Yekutieli D: **The control of the false discovery rate in multiple testing under dependency.** *Annals of statistics* 2001, 29:1165–1188.

[175] Yekutieli D, Benjamini Y: **Resampling-based false discovery rate controlling multiple test procedures for correlated test statistics.** *Journal of Statistical Planning and Inference* 1999, 82:171–196.

[176] Pounds SB: **Estimation and control of multiple testing error rates for microarray studies.** *Brief Bioinform* 2006, 7:25–36.

[177] Reiner A, Yekutieli D, Benjamini Y: **Identifying differentially expressed genes using false discovery rate controlling procedures.** *Bioinformatics* 2003, 19:368–375.

[178] Storey J: **A direct approach to false discovery rates.** *Journal of The Royal Statistical Society Series B* 2002, 64:479–498.

[179] Storey JD, Tibshirani R: **Statistical significance for genomewide studies.** *Proc Natl Acad Sci U S A* 2003, 100:9440–9445.

[180] Pawitan Y, Michiels S, Koscielny S, Gusnanto A, Ploner A: **False discovery rate, sensitivity and sample size for microarray studies.** *Bioinformatics* 2005, 21:3017–3024.

[181] Smet FD, Moreau Y, Engelen K, Timmerman D, Vergote I, Moor BD: **Balancing false positives and false negatives for the detection of differential expression in malignancies.** *Br J Cancer* 2004, 91:1160–1165.

[182] Díaz-Uriarte R, de Andrés SA: **Gene selection and classification of microarray data using random forest.** *BMC Bioinformatics* 2006, 7:3.

[183] Michiels S, Koscielny S, Hill C: **Prediction of cancer outcome with microarrays: a multiple random validation strategy.** *Lancet* 2005, 365:488–492.

[184] Ruschhaupt M, Huber W, Poustka A, Mansmann U: **A compendium to ensure computational reproducibility in high-dimensional classification tasks.** *Stat Appl Genet Mol Biol* 2004, 3:Article37.

[185] Tibshirani R, Hastie T, Narasimhan B, Chu G: **Diagnosis of multiple cancer types by shrunken centroids of gene expression.** *Proc Natl Acad Sci U S A* 2002, 99:6567–6572.

[186] Amaratunga D, Cabrera J: *Exploration and Analysis of DNA Microarray and Protein Array Data.* Wiley-Interscience, 2004.

[187] Dudoit S, Fridlyand J: **Comparison of discrimination methods for the classification of tumors using gene expression data.** *Journal of the American Statistical Association* 2002, 97:77–87.

[188] Hastie T, Buja A, Tibshirani R: **Penalized discriminant analysis.** *Annals of statistics* 1995, 23:73–102.

[189] Fix E, Hodges J: **Discriminatory analysis - nonparametric discrimination.** Technical Report 11, USAF School of Aviation Medicine, Randolph Field, Texas, 1951.

[190] Wennmalm K: *Analytical Strategies for Identifying Relevant Phenotypes in Microarray Data.*. Ph.D. thesis, Karolinska Institute, 2007.

[191] Jain A, Chandrasekaran B: **Dimensionality and sample size considerations in pattern recognition practice.** *Handbook of Statistics* 1982, 2:835–855.

[192] Peduzzi P, Concato J, Kemper E, Holford TR, Feinstein AR: **A simulation study of the number of events per variable in logistic regression analysis.** *J Clin Epidemiol* 1996, 49:1373–1379.

[193] Hastie T, Tibshirani R, Botstein D, Brown P: **Supervised harvesting of expression trees.** *Genome Biology* 2001, 2:0003.1–0003.12.

[194] Bishop C: *Neural networks for pattern recognition.* Oxford University Press, USA, 1995.

[195] Khan J, Wei JS, Ringnér M, Saal LH, Ladanyi M, Westermann F, Berthold F, Schwab M, Antonescu CR, Peterson C, Meltzer PS: **Classification and diagnostic prediction of cancers using gene expression profiling and artificial neural networks.** *Nat Med* 2001, 7:673–679.

[196] Cortes C, Vapnik V: **Support-vector networks.** *Machine Learning* 1995, 20:273–297.

[197] Brown MP, Grundy WN, Lin D, Cristianini N, Sugnet CW, Furey TS, Ares M, Haussler D: **Knowledge-based analysis of microarray gene expression data by using support vector machines.** *Proc Natl Acad Sci U S A* 2000, 97:262–267.

[198] Breiman L: **Bagging predictors.** *Machine Learning* 1996, 24:123–140.

[199] Efron B, Tibshirani R: *An Introduction to the Bootstrap*, volume 57. London: Chapman & Hall, 1993.

[200] Freund Y, Schapire RE: **Experiments with a new boosting algorithm.** In *Proceedings of the Thirteenth International Conference on Machine Learning.* 1996, 148–156.

[201] Breiman L: **Random Forests.** *Machine Learning* 2001, 45:5–32.

[202] Buhlmann P: *Handbook of Computational Statistics: Concepts and Methods*. Birkhäuser, 2004, 877–907.

[203] Ramaswamy S, Tamayo P, Rifkin R, Mukherjee S, Yeang CH, Angelo M, Ladd C, Reich M, Latulippe E, Mesirov JP, Poggio T, Gerald W, Loda M, Lander ES, Golub TR: **Multiclass cancer diagnosis using tumor gene expression signatures.** *Proc Natl Acad Sci U S A* 2001, 98:15149–15154.

[204] Beer DG, Kardia SLR, Huang CC, Giordano TJ, Levin AM, Misek DE, Lin L, Chen G, Gharib TG, Thomas DG, Lizyness ML, Kuick R, Hayasaka S, Taylor JMG, Iannettoni MD, Orringer MB, Hanash S: **Gene-expression profiles predict survival of patients with lung adenocarcinoma.** *Nat Med* 2002, 8:816–824.

[205] Subramanian A, Tamayo P, Mootha VK, Mukherjee S, Ebert BL, Gillette MA, Paulovich A, Pomeroy SL, Golub TR, Lander ES, Mesirov JP: **Gene set enrichment analysis: a knowledge-based approach for interpreting genome-wide expression profiles.** *Proc Natl Acad Sci U S A* 2005, 102:15545–15550.

[206] Raghavan N, Bondt AMIMD, Talloen W, Moechars D, Göhlmann HWH, Amaratunga D: **The high-level similarity of some disparate gene expression measures.** *Bioinformatics* 2007, 23:3032–3038.

[207] Khatri P, Draghici S: **Ontological analysis of gene expression data: current tools, limitations, and open problems.** *Bioinformatics* 2005, 21:3587–3595.

[208] Rivals I, Personnaz L, Taing L, Potier MC: **Enrichment or depletion of a GO category within a class of genes: which test?** *Bioinformatics* 2007, 23:401–407.

[209] Alexa A, Rahnenführer J, Lengauer T: **Improved scoring of functional groups from gene expression data by decorrelating GO graph structure.** *Bioinformatics* 2006, 22:1600–1607.

[210] Falcon S, Gentleman R: **Using GOstats to test gene lists for GO term association.** *Bioinformatics* 2007, 23:257–258.

[211] Pavlidis P, Qin J, Arango V, Mann JJ, Sibille E: **Using the gene ontology for microarray data mining: a comparison of methods and application to age effects in human prefrontal cortex.** *Neurochem Res* 2004, 29:1213–1222.

[212] Raghavan N, Amaratunga D, Cabrera J, Nie A, Qin J, McMillian M: **On methods for gene function scoring as a means of facilitating the interpretation of microarray results.** *J Comput Biol* 2006, 13:798–809.

[213] Nam D, Kim SY: **Gene-set approach for expression pattern analysis.** *Brief Bioinform* 2008, 9:189–197.

[214] Allison DB, Cui X, Page GP, Sabripour M: **Microarray data analysis: from disarray to consolidation and consensus.** *Nat Rev Genet* 2006, 7:55–65.

[215] Goeman JJ, van de Geer SA, de Kort F, van Houwelingen HC: **A global test for groups of genes: testing association with a clinical outcome.** *Bioinformatics* 2004, 20:93–99.

[216] Mansmann U, Meister R: **Testing differential gene expression in functional groups. Goeman's global test versus an ANCOVA approach.** *Methods Inf Med* 2005, 44:449–453.

[217] Tomfohr J, Lu J, Kepler TB: **Pathway level analysis of gene expression using singular value decomposition.** *BMC Bioinformatics* 2005, 6:225.

[218] Efron B, Tibshirani R: **On testing the significance of sets of genes.** *Ann Appl Stat* 2007, 1:107–129.

[219] Damian D, Gorfine M: **Statistical concerns about the GSEA procedure.** *Nat Genet* 2004, 36:663.

[220] Kanehisa M, Goto S: **KEGG: kyoto encyclopedia of genes and genomes.** *Nucleic Acids Res* 2000, 28:27–30.

[221] KEGG: **Kyoto Encyclopedia of Genes and Genomes.** Internet. URL http://www.genome.ad.jp/kegg/.

[222] Ideker T, Ozier O, Schwikowski B, Siegel AF: **Discovering regulatory and signalling circuits in molecular interaction networks.** *Bioinformatics* 2002, 18 Suppl 1:S233–S240.

[223] Nacu S, Critchley-Thorne R, Lee P, Holmes S: **Gene expression network analysis and applications to immunology.** *Bioinformatics* 2007, 23:850–858.

[224] D'haeseleer P: **Data requirements for inferring genetic networks from expression data.** In *Pacific Symposium on Biocomputing '99*. Mauna Lani, Hawaii, 1999, poster session.

[225] Fishel I, Kaufman A, Ruppin E: **Meta-analysis of gene expression data: a predictor-based approach.** *Bioinformatics* 2007, 23:1599–1606.

[226] Bhattacharjee A, Richards WG, Staunton J, Li C, Monti S, Vasa P, Ladd C, Beheshti J, Bueno R, Gillette M, Loda M, Weber G, Mark EJ, Lander ES, Wong W, Johnson BE, Golub TR, Sugarbaker DJ, Meyerson M: **Classification of human lung carcinomas by mRNA expression profiling reveals distinct adenocarcinoma subclasses.** *Proc Natl Acad Sci U S A* 2001, 98:13790–13795.

[227] Chan SK, Griffith OL, Tai IT, Jones SJM: **Meta-analysis of colorectal cancer gene expression profiling studies identifies consistently reported candidate biomarkers.** *Cancer Epidemiol Biomarkers Prev* 2008, 17:543–552.

[228] Wirapati P, Sotiriou C, Kunkel S, Farmer P, Pradervand S, Haibe-Kains B, Desmedt C, Ignatiadis M, Sengstag T, Schütz F, Goldstein D, Piccart M, Delorenzi M: **Meta-analysis of gene expression profiles in breast cancer: toward a unified understanding of breast cancer subtyping and prognosis signatures.** *Breast Cancer Res* 2008, 10:R65.

[229] Toedling J, Schmeier S, Heinig M, Georgi B, Roepcke S: **MACAT–microarray chromosome analysis tool.** *Bioinformatics* 2005, 21:2112–2113.

[230] Pinkel D, Albertson DG: **Array comparative genomic hybridization and its applications in cancer.** *Nat Genet* 2005, 37 Suppl:S11–S17.

[231] Caron H, van Schaik B, van der Mee M, Baas F, Riggins G, van Sluis P, Hermus MC, van Asperen R, Boon K, Voûte PA, Heisterkamp S, van Kampen A, Versteeg R: **The human transcriptome map: clustering of highly expressed genes in chromosomal domains.** *Science* 2001, 291:1289–1292.

[232] Venn J: **On the diagrammatic and mechanical representation of propositions and reasonings.** *The London, Edinburgh, and Dublin Philosophical Magazine and Journal of Science* 1880, 9:1–18.

[233] Ploner A, Calza S, Gusnanto A, Pawitan Y: **Multidimensional local false discovery rate for microarray studies.** *Bioinformatics* 2006, 22:556–565.

[234] Storey J: **The positive false discovery rate: A Bayesian interpretation and the q-value.** *Annals of statistics* 2003, 31:2013–2035.

[235] Hintze JL, Nelson RD: **Violin plots: A box plot-density trace synergism.** *The American Statistician* 1998, 52(2):181–84.

[236] Bitton DA, Okoniewski MJ, Connolly Y, Miller CJ: **Exon level integration of proteomics and microarray data.** *BMC Bioinformatics* 2008, 9:118.

[237] Parkinson H, Kapushesky M, Kolesnikov N, Rustici G, Shojatalab M, Abeygunawardena N, Berube H, Dylag M, Emam I, Farne A, Holloway E, Lukk M, Malone J, Mani R, Pilicheva E, Rayner TF, Rezwan F, Sharma A, Williams E, Bradley XZ, Adamusiak T, Brandizi M, Burdett T, Coulson R, Krestyaninova M, Kurnosov P, Maguire E, Neogi SG, Rocca-Serra P, Sansone SA, Sklyar N, Zhao M, Sarkans U, Brazma A:

ArrayExpress update–from an archive of functional genomics experiments to the atlas of gene expression. *Nucleic Acids Res* 2009, 37:D868–D872.

[238] Barrett T, Suzek TO, Troup DB, Wilhite SE, Ngau WC, Ledoux P, Rudnev D, Lash AE, Fujibuchi W, Edgar R: **NCBI GEO: mining millions of expression profiles–database and tools.** *Nucleic Acids Res* 2005, 33:D562–D566.

[239] Geyer C: **Reproducible Research,** 2008. URL http://www.stat.umn.edu/~charlie/Sweave/.

[240] FDA: **Critical Path Initiative.** URL http://www.fda.gov/oc/initiatives/criticalpath/.

[241] Hiroi N, Martín AB, Grande C, Alberti I, Rivera A, Moratalla R: **Molecular dissection of dopamine receptor signaling.** *J Chem Neuroanat* 2002, 23:237–242.

[242] Dressman HK, Hans C, Bild A, Olson JA, Rosen E, Marcom PK, Liotcheva VB, Jones EL, Vujaskovic Z, Marks J, Dewhirst MW, West M, Nevins JR, Blackwell K: **Gene expression profiles of multiple breast cancer phenotypes and response to neoadjuvant chemotherapy.** *Clin Cancer Res* 2006, 12:819–826.

[243] Huang F, Reeves K, Han X, Fairchild C, Platero S, Wong TW, Lee F, Shaw P, Clark E: **Identification of candidate molecular markers predicting sensitivity in solid tumors to dasatinib: rationale for patient selection.** *Cancer Res* 2007, 67:2226–2238.

[244] PredTox I: Internet. URL www.innomed-predtox.com.

[245] Nutt CL, Mani DR, Betensky RA, Tamayo P, Cairncross JG, Ladd C, Pohl U, Hartmann C, McLaughlin ME, Batchelor TT, Black PM, von Deimling A, Pomeroy SL, Golub TR, Louis DN: **Gene expression-based classification of malignant gliomas correlates better with survival than histological classification.** *Cancer Res* 2003, 63:1602–1607.

[246] Buyse M, Loi S, van't Veer L, Viale G, Delorenzi M, Glas AM, d'Assignies MS, Bergh J, Lidereau R, Ellis P, Harris A, Bogaerts J, Therasse P, Floore A, Amakrane M, Piette F, Rutgers E, Sotiriou C, Cardoso F, Piccart MJ, Consortium TRANSBIG: **Validation and clinical utility of a 70-gene prognostic signature for women with node-negative breast cancer.** *J Natl Cancer Inst* 2006, 98:1183–1192.

[247] Paik S, Shak S, Tang G, Kim C, Baker J, Cronin M, Baehner FL, Walker MG, Watson D, Park T, Hiller W, Fisher ER, Wickerham DL, Bryant J, Wolmark N: **A multigene assay to predict recurrence of**

tamoxifen-treated, node-negative breast cancer. *N Engl J Med* 2004, 351:2817–2826.

[248] Bryce Harrington NHTG MenTaLguY, the Inkscape developers: **Inkscape: The open source scalable vector graphics editor.** URL http://www.inkscape.org/.

[249] Mattis P, Kimball S, the GIMP developers: **The GIMP: The GNU Image Manipulation Program.** URL http://gimp.org.

[250] Gentleman RC, Carey VJ, Bates DM, Bolstad B, Dettling M, Dudoit S, Ellis B, Gautier L, Ge Y, Gentry J, Hornik K, Hothorn T, Huber W, Iacus S, Irizarry R, Leisch F, Li C, Maechler M, Rossini AJ, Sawitzki G, Smith C, Smyth G, Tierney L, Yang JYH, Zhang J: **Bioconductor: open software development for computational biology and bioinformatics.** *Genome Biol* 2004, 5:R80.

[251] Leisch F: *Compstat 2002 - Proceedings in Computational Statistics.* Physica Verlag, Heidelberg, 2002, 575–580.

[252] Gentleman R, Lang T, et al.: **Statistical analyses and reproducible research.** *Bioconductor Project Working Papers* 2004, 2.

[253] Scott T: **Reproducible research with R, LATEX, & Sweave.** Department of Biostatistics Seminar/Workshop Series, 2008.

[254] Garraway LA, Widlund HR, Rubin MA, Getz G, Berger AJ, Ramaswamy S, Beroukhim R, Milner DA, Granter SR, Du J, Lee C, Wagner SN, Li C, Golub TR, Rimm DL, Meyerson ML, Fisher DE, Sellers WR: **Integrative genomic analyses identify MITF as a lineage survival oncogene amplified in malignant melanoma.** *Nature* 2005, 436:117–122.

[255] Glazer MI, Fidanza JA, McGall GH, Trulson MO, Forman JE, Frank CW: **Kinetics of oligonucleotide hybridization to DNA probe arrays on high-capacity porous silica substrates.** *Biophys J* 2007, 93:1661–1676.

[256] Shendure J: **The beginning of the end for microarrays?** *Nat Methods* 2008, 5:585–587.

[257] Margulies M, Egholm M, Altman WE, Attiya S, Bader JS, Bemben LA, Berka J, Braverman MS, Chen YJ, Chen Z, Dewell SB, Du L, Fierro JM, Gomes XV, Godwin BC, He W, Helgesen S, Ho CH, Ho CH, Irzyk GP, Jando SC, Alenquer MLI, Jarvie TP, Jirage KB, Kim JB, Knight JR, Lanza JR, Leamon JH, Lefkowitz SM, Lei M, Li J, Lohman KL, Lu H, Makhijani VB, McDade KE, McKenna MP, Myers EW, Nickerson E, Nobile JR, Plant R, Puc BP, Ronan MT, Roth GT, Sarkis GJ, Simons JF, Simpson JW, Srinivasan M, Tartaro KR, Tomasz A, Vogt KA, Volkmer GA, Wang SH, Wang Y, Weiner MP, Yu P, Begley

RF, Rothberg JM: **Genome sequencing in microfabricated high-density picolitre reactors.** *Nature* 2005, 437:376–380.

[258] Erlich Y, Mitra PP, delaBastide M, McCombie WR, Hannon GJ: **Alta-Cyclic: a self-optimizing base caller for next-generation sequencing.** *Nat Methods* 2008, 5:679–682.

[259] Hillier LW, Marth GT, Quinlan AR, Dooling D, Fewell G, Barnett D, Fox P, Glasscock JI, Hickenbotham M, Huang W, Magrini VJ, Richt RJ, Sander SN, Stewart DA, Stromberg M, Tsung EF, Wylie T, Schedl T, Wilson RK, Mardis ER: **Whole-genome sequencing and variant discovery in C. elegans.** *Nat Methods* 2008, 5:183–188.

[260] Dolan PC, Denver DR: **TileQC: a system for tile-based quality control of Solexa data.** *BMC Bioinformatics* 2008, 9:250.

[261] Huang W, Marth G: **EagleView: A genome assembly viewer for next-generation sequencing technologies.** *Genome Res* 2008, 18:1538–43.

[262] Andries K, Verhasselt P, Guillemont J, Göhlmann HWH, Neefs JM, Winkler H, Gestel JV, Timmerman P, Zhu M, Lee E, Williams P, de Chaffoy D, Huitric E, Hoffner S, Cambau E, Truffot-Pernot C, Lounis N, Jarlier V: **A diarylquinoline drug active on the ATP synthase of Mycobacterium tuberculosis.** *Science* 2005, 307:223–227.

[263] Editorial: **Byte-ing off more than you can chew.** *Nat Methods* 2008, 5:577.

[264] 't Hoen PAC, Ariyurek Y, Thygesen HH, Vreugdenhil E, Vossen RHAM, de Menezes RX, Boer JM, van Ommen GJB, den Dunnen JT: **Deep sequencing-based expression analysis shows major advances in robustness, resolution and inter-lab portability over five microarray platforms.** *Nucleic Acids Res* 2008, 36:e141.

Index